HISTORY, PHILOSOPHY AND SOCIOLOGY OF SCIENCE

Classics, Staples and Precursors

HISTORY, PHILOSOPHY AND SOCIOLOGY OF SCIENCE

Classics, Staples and Precursors

Selected By

YEHUDA ELKANA
ROBERT K. MERTON
ARNOLD THACKRAY
HARRIET ZUCKERMAN

SCIENCE AND EVERYDAY LIFE

by

J. B. S. HALDANE

ARNO PRESS

A New York Times Company

New York – 1975

Reprint Edition 1975 by Arno Press Inc.

Reprinted by permission of
Macmillan Publishing Co., Inc., and
Lawrence & Wishart Ltd.

Reprinted from a copy in
The Newark Public Library

HISTORY, PHILOSOPHY AND SOCIOLOGY OF SCIENCE:
Classics, Staples and Precursors
ISBN for complete set: 0-405-06575-2
See last pages of this volume for titles.

Manufactured in the United States of America

Library of Congress Cataloging in Publication Data

Haldane, John Burdon Sanderson, 1892-1964.
Science and everyday life.

(History, philosophy, and sociology of science)
Reprint of the ed. published by Macmillan, New York.
1. Science--Addresses, essays, lectures. I. Title.
Q160.H34 1975 508'.1 74-26267
ISBN 0-405-06595-7

SCIENCE AND EVERYDAY LIFE

THE MACMILLAN COMPANY
NEW YORK · BOSTON · CHICAGO · DALLAS
ATLANTA · SAN FRANCISCO

SCIENCE AND EVERYDAY LIFE

by

J. B. S. HALDANE, F.R.S.

Professor of Biometry
in the University of London

New York
THE MACMILLAN COMPANY
1940

Printed in the United States of America

CONTENTS

CONTENTS

PREFACE

THE seventy articles collected in this volume deal with various aspects of science. It is often said that modern science cannot be explained in anything less than a whole volume, and that short articles on it are necessarily worthless. I do not agree. The only subjects which are definitely unsuited are organic chemistry, mathematics, and those branches of science which use a lot of mathematics. These have their own terminology, and one cannot even explain in a thousand words what "β-alanyl-histidine" or "an almost periodic function" mean, let alone deal with recent work on them. But a great deal of work in other branches of science is quite easy to follow, at least partially. I can explain that a lot of very small stars called "white dwarfs" are being found fairly close (on an astronomical scale) to our sun. Some astronomers will say that such knowledge is useless and superficial unless I explain about parallaxes, spectroscopic measures of surface temperature, and so on. My answer is that if my critics will tell me just how their boots were made, I will agree. But I don't expect the astronomer to know the details of tanning before he talks of boots, nor need he expect the ordinary man to know the details of astrophysics before he talks about stars.

The ordinary man must know something about various branches of science, for the same reason that the astronomer, even if his eyes are fixed on higher things, must know about boots. The reason is that these matters affect his daily life. In each article of this book I have tried to do two things. I have tried to give a few facts which are not yet to be found in textbooks, and which a student leaving a university with an honours degree would not be expected to know. And I have tried to bring these facts into relation with everyday life. Undoubtedly this is what the ordinary teacher finds hardest, and for a very simple reason. A hundred years ago many books on science stressed the applications of it which were then being made. This is harder to-day because a great deal of our knowledge finds no application in practical life. Whether this is the fault of scientists or of society in general is a very important question.

In the nineteenth century distinguished men of science wrote regularly in the daily Press. Thus Ray Lankester was a regular contributor to the *Daily Telegraph*. This has ceased to be the case because the readers of most newspapers are not interested in such matters, or so at least their editors say. Some sixteen years ago a colleague and I offered to do a free scientific news service for the *Daily Herald*. The offer was turned down. Personally I do not think that the rank and file of the Labour Movement are as stupid as the editorial board of that journal appeared to believe.

Fortunately, however, the *Daily Worker* has now given me the opportunity to write them an article which appears every Thursday, and this book represents my first sixteen months' output. A number of the articles were topical. Indeed that on the fossil fish *Latimeria* appeared two days before the much fuller account in *Nature*. Many of them have also appeared in American and Canadian journals. I must apologise to overseas readers for taking most of my examples from Britain. But I hope that I may encourage writers with a knowledge of local conditions to follow my example in other countries. I am convinced that it is the duty of those scientists who have a gift for writing to make their subject intelligible to the ordinary man and woman. Without a much broader knowledge of science, democracy cannot be effective in an age when science affects all our lives continually. I hope that my book is a contribution to this essential need.

MEALS

BREAD AND HAM

I WAS recently asked to tell a number of school science teachers how to make biology interesting to children. I suggested that they should begin with familiar things, and that breakfast would not be a bad start. A lot of our ignorance and unhappiness comes from taking familiar things for granted. Scientists were not content with observing that the sun rises. They wanted to know why, and found out that the earth is round and spins on its axis. So geography became possible. Socialists were not content with observing how capitalist economics work or don't work. They found out how they had arisen and how they will be superseded.

So with breakfast. Every kind of food has a long history, and will doubtless go on having a history. Let us begin with bread. Bread is two stages removed from Nature. For bread is made from wheat. Wheat is as much a human product as a motor-bus. Certainly our ancestors did not make it out of nothing any more than we make motor-buses out of nothing. But they made it from grasses with a seed yield very much smaller than modern wheat.

There are nearly 300 "natural orders" of flowering

plants, and one of them, the Gramineæ, or grasses, is as important to man as all the others put together. Our bread comes from one; much of our sugar from another, the bamboo cane; our meat and milk from animals which have eaten others; and even our eggs consist to a considerable extent of maize, a grass seed, transformed by hens. Some of us drink still another kind of grass seed transformed into beer or whisky.

The grasses are one of the peaks of plant evolution, not because of their size or complicated structure, but because of the good start in life which they give their children. Instead of producing millions of seeds, of which only a very few germinate, they produce relatively few, with plenty of reserve material, just as a bird, compared with a fish, lays few but well-provided eggs. Man has found these seeds particularly useful as food because they contain plenty of protein, and that of a kind much more suitable for building the human body than the protein of peas and beans.

We do not know the names of the prehistoric men and women who first domesticated wheat. But we do know that of Saunders, who deliberately made Marquis, the famous Canadian wheat, by crossing together a variety which had a high yield with one suitable for growing where spring frosts are late and autumn frosts early. Saunders was a real empire-builder, who pushed the limits of agriculture in Canada northwards. Perhaps some day, when Canadians give honour where it is due, Alberta will be called Saunderia, for I hardly think that Queen Victoria's husband did much to settle the prairie provinces of Canada.

There is another plant concerned in most bread-making —namely, yeast. This is a single-celled plant, and the different bread yeasts are, of course, quite as different from those concerned in making beer or whisky as wheat from barley.[1] Once again we do not know when yeasts were first domesticated, though the Jewish Passover rites seem to show that it was before the Egyptian General Strike which they commemorate. But it was only in the nineteenth century that Hansen isolated single cells of yeast, and established pure lines of it, like the named varieties of wheat.

Just as we import wheat from Canada, we import a great deal of yeast from Holland, where different varieties are grown with as great care as tulips or hyacinths. Yeast is only one of a number of domesticated fungi, including the cultivated mushroom, which differs from the field variety, and the various moulds which make the different sorts of cheese.

Along with bread, you may have had bacon or ham for breakfast. Even if it was Danish bacon, it was very probably of British origin in the past. For about half the pigs in Denmark are large whites, an English breed which probably originated in Yorkshire, whereas the middle whites were formed in England by crossing large whites and small Chinese pigs. If the Danes manage to beat our farmers they do so through co-operation, and not because they have better pigs.

Why, by the way, are bacon and ham red after cooking? This question was first answered by my late father.

[1] This does not mean that a baker's yeast cannot make alcohol under suitable conditions. It can, but not so quickly as a brewer's yeast.

He had shown that deaths in colliery explosions are mainly due to the combination of carbon monoxide with hæmoglobin, the red pigment of blood. After death from other causes, the dying cells use up the oxygen in the blood, so the hæmoglobin becomes a bluish colour. But a corpse remains pink after poisoning with carbon monoxide.

Hæmoglobin forms a pink compound with another gas, nitric oxide, and very occasionally people are poisoned by nitrites, and die for the same reason as in coal-gas poisoning, because their blood will not carry oxygen round. When bacon and ham are salted with saltpetre (potassium nitrate) some of the salt is converted to nitrite, and this gives a red compound with hæmoglobin so stable that it is only slightly altered by cooking. But meat preserved by common salt or smoking is brown when cooked because it contains ordinary hæmoglobin, which can combine with oxygen or give it up again. Just because it is useful in this way it is a far less stable stuff than the nitric oxide compound, and is destroyed by cooking.

BUTTER AND EGGS

THE biologist sees a very fundamental difference between bacon and eggs. To get bacon, you have to kill the pig. To get eggs, you must keep the hen alive. It is true that you kill an egg when you boil it, but a great many eggs are laid by hens which have not been near a rooster for some time, and therefore could not possibly develop into chickens.

Now, poultry were not originally domesticated as egg-layers. They were kept to be eaten, and laid very few and small eggs. This is still the case in India, where the wild jungle-fowl was probably first domesticated. The modern hen is a product of artificial evolution. And this evolution has occurred mainly in England, France, Holland and Belgium, though some breeds, such as the Wyandotte and Australorp, have been produced in the United States and the Dominions.

The evolution of the hen is an example of the law, first pointed out by a group of Soviet biologists, that a domesticated animal generally passes through several stages of exploitation. In the first stage it is of no use till it is killed. The animal is merely kept to save the trouble of hunting it. In the second stage it is used while alive, and also killed. In the third stage it is only used when alive.

The pig is still entirely in the first stage. Pigs have been selected for rapid growth and high fertility, and for little else. Beef cattle, such as the Aberdeen Angus, are also in the first stage. But as soon as cattle are bred for milk production or for pulling carts or ploughs, they pass into the second stage. And in England, where we eat very little horseflesh, the horse is in the third stage.

There are, of course, exceptions to this rule, notably the dog, cat, canary and bee, which were first domesticated for their services and not for their meat or hides, even if dogs are eaten in China and cats sometimes skinned for their fur. Nevertheless, it seems to be a general law. At the present moment the rabbit is passing from the first to the second stage. Most domestic rabbits are only used for their meat and skins, but the long-haired Angora rabbit is kept, like a sheep, as a source of wool, and yields several crops in the course of its life.

Whereas we can get wool from both male and female sheep, and work from both male and female horses, in poultry and cattle only the females give eggs or milk, and most of the males are killed and eaten early. So these animals will only reach the second stage of domestication when we find out how to control their sex, and produce no more males than are needed to carry on the race.

This is not impossible. So far injections of female sex hormones into eggs have turned chickens which should have been cockerels into intermediate types. There is no reason why they should not be switched over completely to pullets when we learn more.

Both the hen and the cow are highly specialised. The pedigree cow with a high milk yield is generally more delicate than the common breeds, but not very much so. On the other hand, the hen with a yield of 300 or more eggs per year is very unhealthy. Great numbers die from disease of the egg-laying apparatus, and, besides this, fowl paralysis and other diseases are spreading and playing havoc with our poultry industry.[1]

[1] I have been violently attacked in the fascist Press for this statement. I may be wrong about fowl paralysis. A distinguished poultry breeder told me he didn't think it had any further to spread in England! But I stick to my statement as to the unhealthiness of the high-laying breeds. It is, of course, true that a hen must remain in health to lay 300 eggs during a year. But the statistics of egg-laying trials prove that these hens have a shorter expectation of life than ordinary barnyard fowls.

It is fairly certain that research would greatly improve matters. But very little is being done. If the poultry industry could put pressure on the Government, a great deal more would be done. Probably the best results would come from pure research on poultry not immediately directed to a practical end. For example, Bateson and Punnett in Cambridge discovered sex-linkage in poultry (which I shall explain in a later article[1]) when investigating the inheritance of colour, which in itself is of very little economic importance.

But Punnett saw that this gave a means of distinguishing the sexes of chicks at hatching. This was so useful that in 1932 one single firm raised 800,000 chicks from crosses involving sex-linkage, and probably about 10 million are now raised yearly in Britain from such crosses. In the same way research on the physiology of egg-laying might give a clue to the combination of high egg yield and good health. However, since the Government expenditure on agricultural research was cut by 38 per cent. this year, it is clear that nothing is likely to be done on these lines.

The butter yield of British cows could be greatly raised by selective breeding. In practice this can only be done on a large scale by using the best bulls. Now we can only judge of the quality of a bull by testing his daughters. In a Danish experiment one bull had sixteen daughters, each of whom produced a higher percentage of butter than her mother. Another had eleven out of twelve daughters worse than their mothers.

So we can only trust to bulls so old that a number of their daughters have already been tested. Mass breeding

[1] p. 257.

from old bulls can be done in the Soviet Union where collective farms are very large and artificial insemination is used, so that one bull had 1,450 calves in a single season. Improvement is bound to be very much slower with small-scale agriculture, as in Britain. England led the world in animal breeding in the eighteenth and nineteenth centuries. Unless we organise our agriculture, we shall lose this lead in the twentieth century.

•

WHY BANANAS HAVE NO PIPS

CHILDREN ask very awkward questions until we grown-ups knock the sense out of their heads and teach them to accept the world as they find it. I am sure that thousands of them have asked their parents why bananas have no pips or stones. For almost every other fruit has some kind of seed, usually in the middle, but sometimes outside, as in the strawberry.

Some parents probably said that God had made the banana like that. This is certainly untrue, for all wild bananas have hard pips about the size of cherry stones, from which new banana trees grow if they are planted. How then did the domestic banana lose its pips?

The answer has only been discovered in the last five years, and throws light, not only on plant-breeding, but on the origin of cultivated plants and of human civilisation.

Plants can be propagated in two distinct ways. They

can be sexually reproduced. In this case, the eggs which develop into seeds are fertilised by pollen either from a special male flower, as in the begonia or the alder, or from the male part of a hermaphrodite flower.

Or they can be propagated without a sexual process, by budding, cutting or grafting. If you buy a tulip, rose or potato belonging to a named variety, this means that it is derived by simple division from one particular seedling. For example, every apple tree of the variety Cox's Orange springs from one apple pip which germinated in a garden at Slough more than a century ago.

This seedling gave such good apples that its twigs have been grafted on to millions of different stocks in every continent of the world. In the same way, all potatoes called Arran Victory and all tulips called Inglescombe Yellow come from a single seed, and are, of course, extremely like one another.

It is a lucky thing that human beings cannot be propagated in this way. If they could we should probably be being ruled by cuttings of King Alfred (the man, not the daffodil) or William the Conqueror! They had the qualities needed for success in their own day. But they might not be well adapted for the very different problems of our own.

This method of propagation is possible even when a plant sets no seeds at all. The tiger lily, many double begonias, and some tulips are quite sterile. So, of course, are seedless oranges and grapes. And so are the bananas which we eat in this country.

One great advantage of asexual propagation by cuttings or grafts is that you can get a very uniform set of plants. Whereas it is much harder to get a plant to breed true from seed. Any commercial apple, tulip or potato gives a great variety of seedlings, mostly much worse plants, from a commercial point of view, than itself.

It takes about ten generations of self-fertilisation to get a seed-propagated plant, such as a wheat or pea, to breed quite true. And it would take a century to make an apple tree all of whose pips, when planted, would give trees like the parent.

Monopoly capitalism demands uniform products. So it naturally favours vegetative propagation of a few standard types of tree. And 95 per cent. of the bananas eaten in England come off trees derived from one single seedling called Gros Michel, or Fat Mike. Millions of these trees are planted on every large West Indian island.

But standardisation, like other features of monopoly capitalism, has its weak side. The roots of Gros Michel are attacked by a particular fungus which causes a wilting of the leaves known as "Panama disease." Many other varieties are more or less immune. So, but for the standardisation, the fungus would have done no great damage to the banana trade.

But the wilt is now spreading through the West Indies, and frantic efforts are being made to breed a banana in which the good qualities of Gros Michel are combined with resistance to Panama disease. This would not be hard in the case of a seed-propagated plant like wheat. One can cross a good cropping variety with one which is immune to rust (a fungus attacking the leaves) and in the second or third generation one generally gets a plant combining both characters.

But Gros Michel is sterile. You can only very rarely get seeds from it, or even use its pollen on a fertile variety of banana. It is, in fact, very nearly a dead end so far as breeding is concerned, though if pollen from a seed-bearing variety is used, about one fertile seed can be obtained from five banana trees.

So the big fruit companies got worried, and the Imperial College of Tropical Agriculture started to cross bananas and to find out why the commercial varieties have no pips. Botanists began to dig up the roots of banana trees and embed them in paraffin wax. They were then cut into sections so fine that the cells could be examined under a powerful microscope.

In the dividing cells they were able to count a number of little bodies called chromosomes, which are the material basis of heredity.[1] And they found that the natural species of banana had twenty-two chromosomes, whilst Gros Michel and the other sterile forms had thirty-three.

Now every cell of an ordinary plant or animal has an even number of chromosomes, half derived from each parent. Thus a man or a woman gets twenty-four from the mother, and the same number from the father. And half of these forty-eight chromosomes go into each sex-cell, whether it is an egg or a spermatozoön, so that the number remains constant in each generation.

An odd number is a sign either of hybridisation between different species or of an accident in development, and it

[1] See p. 244.

always causes sterility. For the chromosomes cannot be evenly distributed to the sex-cells. Thus Gros Michel produces pollen grains of many different sizes, containing a variable number of chromosomes, and useless for fertilisation. A mule is sterile for the same reason. It has a set of horse chromosomes and a set of donkey chromosomes, and cannot divide them evenly.

So a number of banana trees with thirty-three chromosomes have been bred in Trinidad. Some of them are immune to Panama disease, but so far none of them have all the other good qualities of Gros Michel. It is no joke trying to breed for sterility, and until now all the seedless bananas which are of any value to man have originated by accident, not design. I shall explain later on the lucky accident which led primitive men to improve the cereals and potato unconsciously.

At present a race is going on between the breeders and the fungus. If the breeders win, we shall be eating a new sort of banana in ten years. If the fungus wins, the price of bananas will go up in England, and there will be unemployment and hunger among the West Indian negroes, and very possibly disturbances like those which occurred last year in Trinidad.

THE BEGINNING OF AGRICULTURE

TO-DAY we are in the middle of a time of great economic change, which is causing our existing society to break up. The main cause of this change has been the introduction of machinery, which, on the one hand, made it possible for every man, woman, and child to enjoy comfort and leisure, but, on the other hand, made it impossible for the individual worker to own his means of production.

So far as we know, there has only been as big a change as this once before in man's past. And that was caused by the domestication of plants and animals during the Neolithic Period, the age of polished stone tools, which came between the Palæolithic Age of chipped stone and the age of metals in which we live.

A hundred years ago people still thought that God had provided Cain, the first tiller of the soil, with wheat and other crop plants ready made. This cannot be true. Wheat can be crossed with several grasses which grow wild in Europe, Asia and Africa, and is clearly derived from one or more of them. In the same way, maize, which came from America, will cross with wild Mexican grasses, but with nothing in the Old World.

And the natives of America found very few animals which could be tamed. In Peru the llama was a poor substitute for the horse, cow and sheep. The bison of North America was too big to tame, and the guinea-pig was not a really good source of meat. In Mexico there was not even the llama as a beast of burden. So men were used for this purpose. The great pyramids of ancient Mexico were made by human labour, and men were sacrificed there in thousands when our own ancestors at a similar stage were feeding their gods on the flesh of animals.

If we are to understand the present, it is much more

important to find out all we can about the origin of agriculture than about who killed King William Rufus of England or King James III of Scotland. This seems fairly obvious, but the first man who attacked the problem scientifically from a Marxist angle was the Russian botanist, Vaviloff, who is in charge of a chain of plant-breeding stations in the Soviet Union.

The Soviets needed potatoes which could be grown in the Arctic and stand up to frost. So an expedition was sent to the Andes in Peru and Bolivia, where the potato was first grown, and where its wild ancestors still grow close to the snow-line in the Andes. They got their plants, and by crossing with ordinary potatoes obtained plants which, though still far from perfect, give a fair crop on the shores of the Arctic Ocean.

At the centre of origin of the potato there are a great many different kinds, both wild and domesticated, but only a small fraction of these have been worth exporting to the rest of the world. And just the same is true of other plants. Vaviloff found that there are twenty varieties of bread wheat in the whole of Europe, fifty-two in Persia and no less than sixty in Afghanistan.

So it became clear that bread wheats originated in or near Afghanistan. On the other hand, macaroni wheat, which does not cross easily with bread wheat, originated in the eastern Mediterranean basin, and some kinds of barley in Abyssinia.

When the results on different plants were put together, it became clear that plants were first domesticated in mountainous regions, where men lived in small communities. Only later did their descendants come down into the great valleys of the Nile, Euphrates and Indus, where irrigation was necessary, and great cities arose with priests, kings and a rigid class system.

We know much less about the origin of domestic animals, because their wild ancestors have often been killed off, while the wild species from which our plants came still survive, sometimes as weeds. Nevertheless, a group of Soviet zoologists is trying to tackle the problem too.

How were domesticated plants improved from wild grasses or herbs to their present state? At one time it was thought that God had made them as they now are, just as kings were supposed to reign by divine right. Then Hobbes and Rousseau thought that primitive men had got together and made a contract to form a State. In the same way, many botanists still think that primitive men were so clever as to be able to pick out the best wheat or potato plants as parents for future generations.

I do not believe this for a moment. The State grew up, probably because the beginnings of the class struggle made it necessary, not because anyone designed it. In the same way, some plants were improved automatically. As soon as wheat, beans or any other seeds are harvested each year and the seeds sown on prepared ground, the plants with most seeds are automatically selected.

Other qualities useful to wild plants cease to count, and are lost. Our domestic plants cannot thrive in competition with wild plants. A neglected field is soon choked with weeds. But if their seeds are gathered and carefully planted in soil where competitors have been killed off by the plough and harrow, they do very well.

In the same way, the ancient Peruvians learned to plant potatoes and automatically found that they were improving the potato by choosing the plants which made the most and biggest tubers. They probably ascribed the improvement to magic.

This process only works when the part of the plant used for propagation, and thus unconsciously selected, is also the part eaten. If not, it works in the wrong way. A seedsman selects a nice-looking grass plant from a meadow, sows its seed, and sows this again for a few generations. He finds that he has selected for seed yield, but often lost the qualities most valuable in a meadow grass. For cows and sheep do not want grass seed, but stems and leaves.

When Professor Stapleton of Aberystwyth discovered this principle he was able to produce very greatly improved grass seed for meadows and pastures; he and his colleagues are now busy with clovers and other pasture plants. Fruit trees and bushes are not improved by any automatic process, and we are only now groping our way towards scientific methods of breeding new kinds.

In the same way, many domestic animals were automatically selected for fertility and rapid maturity. Once men had learned to protect wheat from weeds and hens from hawks, they began to increase their yield. A denser population became possible. A man could own a herd which gave him more meat, hides and milk than he needed. He had the capital needed to extract surplus value from the labour power of a slave or a hired man. Primitive communism was doomed, and the distinction of classes based on the ownership of property had begun.

CHRISTMAS, THE DAY AND THE DINNER

DURING its first four centuries, the Christian Church was uncertain as to the date of Jesus' birth. It was sometimes celebrated on January 6th, March 25th or even in November or May. But December 25th was kept as a holy day in England and other countries before they became Christian. The worshippers of Mithras celebrated it as the birthday of the Unconquered Sun.

For it is about Christmas-time that the days begin to lengthen perceptibly. The shortest day is December 22nd, but this could not be ascertained until there were accurate clocks, probably not till the seventeenth century. So the date was found in another way. As the days shorten in autumn, the sun rises and sets further south each day, and the farthest south is on December 22nd.

This was discovered a long time ago. A "wise man," probably a priest, could stand at some definite point, such as the altar stone in the middle of Stonehenge, and mark the directions of sunrise and sunset on each clear day. One of the large stones outside the main circles at Stonehenge marks the direction of sunrise on Midsummer Day. The days when the sun sets furthest north and south, June 22nd and December 22nd, are called "solstices," or sun-stops.

When it was found that they occurred at intervals of 365 or 366 days, there was a basis for a calendar, and knowledge of the calendar was one of the ways in which the priests gained power. This is an example of primitive science, by which men are able to predict what will happen. The second stage of science is the experimental stage, in which men can control events. But this stage only arose when man learned to work so accurately that an experiment can be repeated with the certainty that the result will be the same.

So the wise men of ancient times, who could calculate, but were ashamed to work with their hands, never made accurate experiments. They often worshipped Nature, instead of trying to control it. The festival at the winter solstice seems to be a bit of old nature religion, which has been taken over by Christianity.

There are also some very modern features in our celebration of Christmas. Those who can afford it, buy a turkey. But turkey was a rarity before the eighteenth century. For the turkey and the guinea-pig are among the few domestic animals which come from America. This continent has given us some very important cultivated plants, such as potatoes, tobacco and maize. But it was very poor in domesticable animals.

This was probably the main reason why civilisation did not develop there so quickly as in the Old World. The llama and alpaca are not so good as the horse, ox or camel at carrying loads, nor so good as the sheep at making wool. Nor are they as good milk-producers as the cow or even the goat. No one has ever been able to domesticate the bison or the moose.

If we try to imagine a civilisation without either machines or animals as sources of power, we shall realise what an uphill task the old Mexicans had, and shall expect great things from their descendants, now that they have regained their national liberty and have also got modern productive forces at their disposal.

At the present moment men are engaged in domesticating a number of American fur-bearing animals—the silver fox, the musk-rat or musquash, the coypu or nutria, and the mink. None of them has yet become tame, but their descendants may be so in a century. The fur industry is now passing from the hunting to the pastoral stage, as the meat industry did thousands of years ago. And since trapping is the cruellest form of hunting, this is a good thing even from the animals' point of view.

The turkey, which is a native of the United States, seems to be easier to rear there than in England as a domestic animal. It is certainly a great deal cheaper. Our traditional Christmas dish is not the turkey, but the goose, which is descended from a wild species native to western Europe. But the goose was one of the good things of Old England which became rare after the commons were enclosed.

The Wiltshire commons used to be covered with great flocks of geese, each flock having the owner's mark punched on the web between its toes. The village goose-herd, or "gozzard," drove them home to their various pens each evening. Besides being eaten and providing eggs, they were plucked for their feathers several times a year.

When the land of England belongs to the people once

more, it is probable that most cattle and sheep will belong to collective farms, as in the Soviet Union. But I hope that we shall go back to our old methods of raising geese. If so, there will be a British goose for every British family at Christmas.

ROUND THE YEAR

SUMMER TIME

WE put our clocks an hour forward last week-end. And everything would have been all right if some children hadn't started asking awkward questions. "Mummy, is it really four o'clock when the clock says five?" "Daddy, do we really get up an hour earlier?" These are very hard questions to answer truthfully. If you say, "It's really four," you are getting perilously close to the false view that the whole world is a sort of giant clock, a mechanical affair. If you say, "The time is what we say it is. There isn't any real time," you are on the way to saying that the world is not real, but only our idea.

The truth lies in between those views, but we can only understand it if we look at it from a historical angle. You might try to explain the meaning of "real" time by saying that when the clock-hands point to 12 noon the sun is at its highest, or that it is just halfway between sunrise and sunset.

But that isn't true. All clocks in England and Scotland are supposed to keep the same time. But the sun is at its highest point at Land's End about half an hour later than at Lowestoft, these being the farthest west and east points in England. And it is eight minutes later still in Barra in the Outer Hebrides.

Still, many people think that at noon by the clock the sun is at its highest at Greenwich. Unfortunately that isn't true either. Judged by the sun, the Astronomer Royal's best clock at Greenwich is fourteen minutes, twenty-five seconds fast on February 11th and sixteen minutes, nineteen seconds slow on November 6th. It is right within a few seconds on April 16th, June 15th, September 1st and Christmas Day. In fact, the Astronomer Royal sets his clock, not by the real sun, but by a fictitious heavenly body called the Mean Sun, which keeps time with the clock, and only agrees with the real sun on the average. That is why the official time is called Greenwich Mean Time.

The history of time-measurement is long and complicated, because there are three different natural periods—the year, the month (meaning the time between one new moon and the next) and the day. Besides these, there are artificial periods, such as our week, in which each day has a name, and the twenty-day period of the Mayas in ancient Mexico, whose days had names such as Eb, Ben and Ix.

The priests who made the first calendars had terrible difficulties in trying to fit the natural periods together. The average length of the month is about twenty-nine and a half days, so twelve do not make a year, but only 354 days. The usual method was to insert an extra month every three years or so. Our present artificial months, which do not correspond to the moon, were fixed less than 2,000 years ago.

But there is no natural period shorter than a day. The ancient Egyptians divided the day and the night into twelve hours each. So the hour was not constant. This did not much matter in Egypt, where the length of the day does not vary much. But it would be useless in Europe. So, finally, the day and night together were divided into twenty-four equal hours.

Clocks were invented in the Middle Ages. They could not have been invented earlier in Europe, because the men who had the needed intelligence and education disdained manual work as only fit for slaves. So clocks were first made in a society where workers owned their own tools individually as members of guilds or collectively as monks. But these clocks, though useful for timing church services, had constantly to be set by the sundial.

In the eighteenth century very accurate clocks were designed for navigation. This is no longer necessary, since ships' chronometers can now be set every day by radio. However, as a result of this economic demand, it became possible to show, what had long been believed on other grounds, that the days, measured from one noon to the next, were not of the same length, being longest in January.

The reason is as follows. The earth rotates on its axis once in twenty-three hours fifty-six minutes. Suppose we fix a telescope so that any particular "fixed" star—say, the first of the seven in the Plough—comes in view at a given hour, it will reappear four minutes earlier the next day. The period between two appearances of the same star is called a "sidereal day," and is very nearly constant, so it is used as the standard for regulating clocks.

But the earth goes round the sun once a year; so the sun seems to move against the background of the stars, lagging about four minutes in each day. Thus there are 366¼ sidereal days in the year, but only 365¼ days of the ordinary kind. Now, if the earth went round the sun uniformly in a circle, all the solar days would be of the same length. But as it is nearer to the sun in January and moving faster, this is not the case.

Hence our standards for accurate time measurements are given by the stars and not the sun. So when accurate time-keeping was needed for navigation, an Astronomer Royal was appointed to regulate the official time and to publish the *Nautical Almanac*. It is not a coincidence that Greenwich Observatory was founded in 1675 and Calcutta in 1686.

In the last twenty years there have been great improvements in clock-making. Modern pendulum clocks swing in a vacuum to avoid air resistance, and it is possible, though not yet quite certain, that they are so accurate as to show that even the sidereal day varies. This is to be expected. Thus if in a warm year a lot of ice melts at the Poles and water therefore flows towards the Equator, the effect should be to slow down the earth's rotation and lengthen the day.

Besides these changes, there are no doubt others which take place more slowly. For example, the tides exert a braking action on the earth, which probably makes each day about one six-hundred-thousand-millionth part longer, on the average, than the day before. This fraction is far too small to measure directly, but can be calculated from ancient records of eclipses.

So Greenwich Mean Time is nothing real. It is a measurement of something real, but, like everything human, it has a history which can only be understood in relation to economic history. There is nothing sacred about it, and perhaps some day a better method of measuring time than either stars or clocks will be invented. The only certain prediction is that we shall never be able to measure time with absolute accuracy.

It is undoubtedly convenient to change our clocks forward in the summer, and there is no good reason against such a change, since Greenwich Time is only a human invention. But though clocks and measurements are human, time is real.

POLLEN

DURING the last week[1] several of my friends have developed hay fever. The country air is full of the pollen of various grasses, and there is enough in the air of most towns to irritate a sensitive nose.

The pollen from a flower consists of grains which can easily be seen with a microscope, and which are produced in the male organs of a flower, and are generally needed to fertilise the female organs. A few flowers, like the dandelion, produce pollen which is completely useless. The seeds are formed without its playing any part.

In fact, the pollen of a dandelion has only a historical

[1] In July, 1938.

significance. There are many functionaries in the English State in much the same position. They may have been useful in the past, but they have long ceased to be so, and could be abolished without loss.

The function of pollen could only be discovered by experiment, and the Greeks and Romans rarely made experiments. Those of them who were interested in science observed and reasoned. But they left any work involving manual skill to slaves. It was only when it ceased to be shameful to work with one's hands that anyone did such a simple experiment as cutting off the stamens of a flower where the pollen is formed.

If the flower is now covered with a paper bag, it will not set seed unless pollen is put on to the female organ from another flower. But sex in plants was only discovered towards the end of the seventeenth century. Some plants commonly fertilise themselves. Wheat, peas, and violets are examples. Others, such as red clover and sweet cherries, cannot do so. In a third class, such as maize, sour cherries, and primroses, self-fertilisation is possible, but harmful.

In the plants which are self-sterile we find a system rather like that which Engels describes in *The Origin of the Family* as the basis of one kind of primitive communism. A tribe is divided up into several clans or *gentes*, often called after animals, and each is a self-governing economic unit. A woman of the Bear clan has to marry a man of another clan, say, the Wolf or Snake clan, but her children belong to her own clan.

But although the husband is a Wolf, his children belong to the mother's clan and are Bears. So when he dies his property does not go to his sons, but is divided up among the Wolves. Hence there is little chance of one family becoming wealthy at the expense of others.

If you plant an orchard with cherry trees all of the same kind, say Early Rivers, you will get very little fruit. For pollen grains do not fertilise flowers of a plant belonging to the same clan. But if trees of different clans are mixed, bees carry pollen from one tree to another, and the pollen grains then burst, and form tubes which grow down into the female part of the flower, and fruit is formed.

So Crane has classified our chief types of cherry into clans, and a properly designed orchard must have trees belonging to several different clans. For example, Bedford Prolific belongs to the same clan as Early Rivers, so its pollen is no good for fertilising the latter. However, the analogy is not complete. If cherry stones are planted, the seedlings never belong to the mother's clan, and seldom to the father's. Crane's experiments, which were carried out in South London, took many years, and were done on plants in a greenhouse protected against insects by wire-gauze windows and doors.

Although pollen grains are so small, they are extraordinarily immune from decay. Cannel coal is largely made of fern spores, which are very similar to pollen grains. And peat or mud laid down in ponds many thousands of years ago contains pollen grains which can be recognised with a microscope. For each plant produces grains of a particular shape.

So a biologist can tell what plants were growing near a particular pond many thousand years ago, and in this way the changes in climate for the last 10,000 years have been followed. After the last Ice Age the first plants were Arctic willows. Then the weather improved, and about 5,000 years ago it was much warmer than now. Hazels grew far to the north of their present limit. About 1800 B.C. the weather got colder again, but there was a change for the better about A.D. 450.

So characteristic are some pollen grains that a detective story might be written, and, for all I know, has been, where the criminal's movements are traced by the pollen grains of a rare plant on his clothes.

Just as the pollen grains of each plant have a special shape, so they have a special chemical composition. Sufferers from hay fever are usually sensitive to only one or two kinds. Very often an extract from these will produce a weal if rubbed into a scratch on the skin. The patient can then tell what plants to avoid, and can sometimes be cured by injections.

Men only use pollen for fertilising plants. But bees eat it, and store it as "bee bread" in special combs apart from the honey. And it may yet find some use in human industry.

KEEPING COOL

It is quite likely that this article will be printed on the coldest August day for several years. But if the weather is seasonable, people will be having some difficulty in keeping cool. Why do we need to keep our temperatures constant? Most other animals do not. A fish can live in the same pond whether it is quite warm or close to freezing point.

But we share the need for a constant temperature with the most recently developed classes of animals—namely, birds and mammals, of which man is one species. And the organ which needs a constant temperature is the brain. Muscles will still work when cooled down a great deal, or our hands would be paralysed whenever they got cold. So will glands, such as the kidney. But a rise of a few degrees in the brain temperature causes the delirium of high fever, and an even smaller drop causes unconsciousness.

Indeed, we can learn from a study of fossil skulls that the brain has only developed greatly in size or complexity in birds and mammals since the time when we believe that they started keeping their temperatures steady. Snakes, lizards, frogs and so on have had just as long as ourselves to evolve a good brain. But they could not do so without temperature regulation.

We have three ways of keeping cool. One is by regulating our environment by opening windows, taking off clothes and so on. This is not as trivial as it sounds. Some mothers forget that their babies cannot do this for themselves. They keep them quite needlessly and even dangerously wrapped up on warm days.

The second way is by increasing the blood supply through the skin in order to lose the heat which we are constantly making in our bodies even when at rest, and still more when working. This heat would be enough to kill a resting man in three or four hours if he were kept in a heat-tight vessel such as a "thermos" flask.

The blood vessels, and particularly the arteries, have their own set of muscles, whose relaxation allows much more blood to pass through them than when they are contracted. These muscles are controlled by nerves, but are not under the control of the will. In this respect they are like the heart and the iris (the blue or brown ring round the pupil of the eye) and the muscles of internal organs, such as the intestine.

However, although we cannot contract these muscles at will, they are controlled by a special part of the brain. And this part can learn and forget its functions. If a man has lain in bed for some weeks and then gets up suddenly he faints, for the blood runs down out of his head because the vasomotor centre, as it is called, in the brain, has lost its knack of contracting vessels in the stomach and elsewhere.

The third line of defence is sweating. But sweat is only useful when it evaporates, a process which uses up a lot of heat. Sweat which drips off one is useless. In very dry air one may sweat terrifically without getting wet; in fact, the main thing to be seen is crystals of salt on the clothes.

Hence the ordinary thermometer does not tell much about the effect of heat on the human body. Dry air at body temperature (98° F.) is not very trying. But completely moist air at this temperature is deadly. The body has no way of losing heat, and steadily warms up.

In England, where the air is never very dry, a temperature of 100° F. is very oppressive, whereas in the dry air of Australia test matches have been played in the sun when the shade temperature was 110° F. For this reason we can learn more about the climate from a wet-bulb thermometer—that is to say, a thermometer with its bulb wrapped in a wet cloth—than from the ordinary dry-bulb kind.

The sweat does not just soak through the skin, but is actively secreted by small glands which are under the control of the brain, but not of the will. Hoofed animals, such as cows and horses, have sweat glands. But many others do not. Dogs only have them on the soles of their feet.

However, a dog keeps cool by evaporation from his mouth and tongue. He can secrete two different sorts of saliva, a sticky kind when eating and a watery kind when hot. He also has two kinds of breathing, a regular kind normally and an irregular kind to cool himself. If you watch a dog who is hot, from external heat and not from exercise, and who therefore needs to cool his tongue, but not to ventilate his lungs more than usual, you will see that most of his breaths are shallow, but about one in ten is deep.

One boy in several million is born with no sweat glands and very poor teeth. Girls never have this abnormality. A boy of this kind lives or lived until recently in South London, and in hot weather has to throw a bucket of cold water over himself from time to time.

The sweat has, of course, to be replaced by drinking water and eating salt. Babies sometimes fall ill and die in hot weather because they do not get enough water to drink. And stokers and miners get cramp for lack of salt. Salt is a necessity, not a luxury, in hot countries, and

I hope that one of the first acts of a self-governing India will be the abolition of the salt tax.

THUNDERSTORMS

MOST primitive peoples attribute thunder to the direct action of a god. Sometimes he is the chief god, like the Roman Jupiter, and sometimes a specialist in thunder, like the Anglo-Saxon Thor, after whom Thursday is called. But quite a small air raid causes more noise and more destruction than 100 thunderstorms. So it is lucky for modern religions that none of them is centred round lightning.

The first man to bring thunderstorms within the domain of natural law was Benjamin Franklin, the great American who combined scientific and revolutionary activities 150 years ago. He showed fairly conclusively that a lightning flash was merely a large electric spark. And lightning conductors were put up to protect buildings. But many are so badly designed as to be almost useless. It has also been known for a long time that the distance of a flash can be estimated from the fact that light travels very quickly, but sound takes nearly five seconds to cover a mile. So the distance of a flash in miles is the number of seconds between lightning and thunder, divided by five.

Very little more about lightning was found out until a few years ago. Then it began to interfere with industry. A lightning flash hitting an electric transmission line may cut off light and power from a whole district. And, of course, thunderstorms cause atmospherics on the radio. Finally, the development of high-tension currents made it possible for the first time to make sparks which began to resemble lightning flashes.

A thunderstorm is generally a local upsurge of warm, moist air from near ground level to a higher level. As it rises, it expands, because there is less air above it to compress it. And therefore it cools. This is the converse of the effect known to every cyclist, by which the air in the pump is heated by compression.

Because the air is cooled, the water vapour in it condenses first into mist and then into rain. But the ascending current, which is strongest at the front of the cloud, prevents the fine drops from falling, though large drops and hailstones can do so. These large drops are found to have a positive electric charge, perhaps due to their friction with the air.

Simpson and Screase have recently invented an apparatus for exploring the electrical state of clouds. This consists of a small balloon with about 70 feet of fine wire trailing below it. If there is an electric field in the cloud, a current flows through the wire and is recorded by a special apparatus.

They find that the top of a thunder-cloud is always positive; the bottom negative with occasional positive patches. This agrees with Wilson's discovery that in a lightning flash the earth is generally, but not always, positive and the cloud negative. The power developed is very great. A cloud emitting a flash every ten seconds develops about a million kilowatts. Battersea Power

Station develops only 52,000 kilowatts when working at full capacity.

But the most striking advances in our knowledge of lightning have been made in South Africa with a camera invented by C. V. Boys. Boys is one of our greatest physicists, but we do not hear much about him, because he does not invent new theories, but new apparatus, generally quite simple. His talents would be more appreciated in a workers' state than they are in Britain.

The camera in its latest form has two lenses. Each takes a picture on a revolving film moving past the lens at about 60 feet per second. Now, if a flash takes one ten-thousandth of a second to travel from cloud to earth, the film will have moved about $\frac{1}{14}$ inch in this time. Hence the image is distorted. So by comparing pictures taken with the two lenses the speed can be determined. A cinema camera would have to take 100,000 pictures per second to reach the same accuracy.

The work was done in South Africa for two reasons. Thunderstorms are very common there. And they are so dangerous to power transmission cables that the work was shared between Schonland and Malan of Capetown University and Collens, an employee of the Victoria Falls and Transvaal Power Company, which transmits electricity from the Zambesi to the Rand.

It turns out that a lightning flash is a complicated affair. First comes a "leader" flash from cloud to earth, consisting of steps each about 50 yards long, and often changing direction at each step. This heats the air and makes it a conductor of electricity. Then a much brighter flash goes from earth to cloud along the same track.

This may be all, but generally there are several more strokes along the same track, each being double.

Fairly satisfactory means have now been designed for protecting high-tension cables. And it is possible that our descendants may learn how to store electrical energy in such a way that we can use thunderstorms as a source of energy to furnish us with heat, light and power.

SEA BATHING

SEA BATHING is still[1] tolerable for ordinary people, but the rivers and ponds, which were warmer than the sea in July, are now too cold for most of us. Both depend in the long run on the sun for their heat, but yet they depend on it in a different way. So a study of them may help us to understand how, for example, human actions are governed by economic conditions, though often in a roundabout way.

If we put a thermometer in a vacuum and expose it to sunlight, its temperature is governed directly by the sun. In a cloudless desert, it depends mainly on the angle at which the sun's rays come down, but also to a small extent on variations in the heat which the sun is giving out, which fluctuates by about 5 per cent. on each side of its average value.

The temperature of this thermometer can be calculated fairly accurately, and its graph is a smooth curve. But as soon as clouds appear the curve becomes irregular, even

[1] At the end of October.

though the average temperature varies regularly. Usually the hottest days are not round June 22nd in the Northern Hemisphere and December 22nd in the Southern, when the sun is highest, the days are longest and the vacuum thermometer hottest, but a month or so later.

This lag is due to the fact that the ground takes some time to warm up. The temperature at various depths in the rock on the Calton Hill at Edinburgh was measured by Forbes about 100 years ago with thermometers buried in the rock. The daily variation of temperature did not reach below a depth of 1 yard. As he went lower down the time of greatest temperature became later, but the annual variation less.

Thus at 6 feet the hottest day was in September, and this was 16° F. hotter than the coldest day in March, and at 24 feet the rock was actually hottest in January, though only 1° hotter than in July. In fact, the hottest day on the surface coincides with the coldest 8 yards underground. The temperatures are exactly out of step.

Naturally enough, the sea takes longer to warm up than the land or a small body of water such as a lake or river. The lag is often about three months, so that round England it is generally warmest in September and coldest in March. For the same reason, islands in the ocean have a late spring and a long autumn as compared with inland places.

Sometimes the hottest day comes before midsummer. This is so in many parts of India, where the sky is clear till May or June, and then the monsoon brings rain and a most welcome drop in the temperature. But there is generally a lag. Of course, in the Southern Hemisphere all this is reversed, and the hottest days are usually in January, the coldest in July. In the tropics there may be two hot periods with cooler ones between them.

Of course, in England the weather is much less closely determined by the time of year than in most countries. A south wind and a cloudy sky may give us warm nights in December, while a clear night with a north wind gives frosts in June. But in Irkutsk a thaw in midwinter and a frost in midsummer are equally impossible.

Now let us look at the same kind of events in the economic field. If I say that marriage is largely determined by economic considerations, I shall be told that, even if a few *bourgeois* marry for money, decent men and women marry for love. That is true; but they put off their marriage if the man is out of work, and marry when the man finds a job.

So, although economic factors do not decide when any particular person will marry, they do determine the number of marriages in a given year. I have just been examining the marriage rate in the United States as compared with the economic condition of the country shown by the number of bankruptcies in a year.

The two run closely parallel. Many bankruptcies mean few marriages, and vice versa, as any Marxist would guess. The only exception to the rule is in the year 1918 to 1919, when the War meant few bankruptcies and also few marriages. But we can ask another question. Is the Marxist theory getting truer or less true as time goes on? In other words, is the marriage rate coming to depend more or less closely on economic conditions?

The answer is clear. Thirty years ago the marriage rate

lagged behind the economic index, the maximum of marriages coming a year or two after the minimum of bankruptcies. Now the two occur in the same year, and the fluctuations in the marriage rate are increasing. The ordinary affairs of life are getting more, not less subordinated to economic crises.

As capitalism develops, men and women are becoming more the slaves of the system. They can only break their chains by breaking the system. In capitalist countries we are like houseless savages exposed to the sun and the frost. The citizens of the Soviet Union are like civilised men in houses where they can regulate the temperature to suit their needs.

FREEZING AND THAWING

DURING the last week[1] we have all seen water changing to ice, and ice turning back again to water. It is such a familiar sight that we often forget how strange it is. We know that water and ice both consist of molecules made up of two atoms of hydrogen and one of oxygen. We understand that the molecules are fixed in place in an ice crystal, and can move about freely in water. So it is natural that, to turn ice into water, energy is needed. Heat is a disorderly motion of the molecules, and we can see why motion must be given to the ice molecules to make them melt.

But if this were all, we should expect ice, when heated,

[1] December, 1938.

to behave as glass does. When glass is heated, it softens gradually. When red hot it can be pulled into threads like treacle, and blown into bubbles which stiffen into bulbs when cooled. When it is white hot it is a liquid, and can be poured out of a tank into a mould.

But with ice, and with most pure chemical compounds, the change from solid to liquid is quite sharp. In fact, organic chemists identify compounds by their melting points. For example, if I think I have made some acetyl-salicylic acid (the pain-killing stuff in aspirin, "Koray" and other drugs), I heat it in an oil bath, and see if it melts at 118° C., as it should. Indeed, it is a general rule in chemistry that we can only identify a substance by changing it, either by heating or cooling it, or by seeing how it interacts with other chemicals.

If we cool down water very carefully in a clean, smooth vessel it is sometimes possible to get it several degrees below freezing point without any ice forming. But we have only to touch it with the tiniest crystal of ice to make it freeze. More accurately, some of it will freeze; and the water molecules which become ice give up so much heat that the temperature rises suddenly to melting point and the rest of the water stays liquid.

Many substances are very hard to crystallise. They remain liquid, or stay dissolved in water or some other fluid, until a crystal of the solid is added. When a substance has once been crystallised in a laboratory, it is generally much easier to prepare the crystals a second time, because some of the solid form is already there in the form of dust. The great German chemist, Bayer, had such a marvellous knack of making compounds crystallise that his students said he carried crystals of every substance in his beard, and had only to wave it over a flask to do the trick!

Molten glass will crystallise if it is cooled down slowly enough, and it is then a useless mass of brittle crystals. And if metals are cooled slowly enough, they also make large crystals, which are much softer than the ordinary forms. Crystals may also form in a metal as the result of strain, and for this reason chains used in hoisting tackle are strengthened from time to time by annealing—that is to say, heating them till the crystals are destroyed.

So, given time enough, almost every substance has a definite temperature at which it melts and freezes, like water. But the temperature depends on its purity. If we dissolve a small amount of anything else in water, whether it is a solid, like salt or sugar, a liquid, like alcohol or glycerine, or even a gas, the water becomes harder to freeze. Thus water containing 1 per cent. of common salt freezes at more than 1° F. below the usual freezing-point.

Water is one of the few substances which expand when they freeze. This is why pipes are burst by frost and why special care must be taken to prevent the radiators of cars and lorries from freezing. To do this, a suitable compound must be added to the water. It must not corrode the metal. It must consist of small molecules, for the lowering of the freezing-point is proportional to the number of molecules added. It must not evaporate, or ordinary alcohol would be used. And it must not grow moulds. The best substance so far found is ethylene glycol. But, though it tastes sweet, it is rather poisonous, and has killed a number of people. So never drink your anti-freeze mixture!

One of the finest tests for a chemical substance is by finding what is called a "mixed melting-point." If the chemist who thinks he has made acetyl-salicylic acid wants further evidence that he is right, he will mix some of this compound with his own product. If they are different, the mixture will melt at a lower temperature; if the melting-point is steady, they are the same substance.

H_2O can change from ice to water and back, from water to steam and back, and also from ice to steam and back. For, of course, snow can evaporate without melting, and frost can form without passing through a liquid stage. But these are not the only sudden and reversible changes which it can undergo. If we vary the pressure as well as the temperature, a lot more can happen.

Ordinary ice is bulkier than water. So if we put on pressure to prevent expansion, we hinder the formation of ice and lower the freezing-point. This goes on till a pressure of 2,200 atmospheres is reached, corresponding to that in a well fourteen miles deep. At this pressure the freezing-point is $-22°$ F. Then the ice suddenly collapses into a form in which the molecules are packed together in a different pattern. This form, called "Ice-3," is less bulky than water, so pressure helps it to freeze, and the freezing-point rises again.

I do not know what the present record is, but not so long ago there were six different forms of ice, each passing over sharply into another at the right temperature and pressure. By sufficiently huge pressure Ice-6 can be kept solid at 175° F., and if I had a few million pounds to make the necessary pressure tanks I would guarantee to produce red-hot ice.

So far none of the five newly discovered forms of ice has had any practical application. And only a very highbrow physicist would say that he was going skating on the Ice-1. But such facts as these should make everyone who thinks scientifically very wary of talking about unchangeable properties of substances, or natures of men. The properties of H_2O vary with temperature and pressure, and "human nature" varies according to the kind of society in which men live.

WEATHER AND HISTORY

Does weather control history? About twenty years ago this was a popular theory. Any theory according to which history is determined by something which we can't control is popular with reactionaries. For it means that people who try to make the world better are dangerous visionaries. However, the theory that race determines history is even more useful. Men are prouder of their race than of their weather. It is easier to believe that Englishmen ought to order Indians about because they belong to a higher race than because they live in a colder climate.

Of course, there is some truth in the theory. If our climate changed greatly, we should have to grow different crops, and a change in the mode of production would lead to other social changes. But if climate alone controlled history, New Zealand would have had much the same history as Britain, and Canada as Northern Russia.

How do we know the weather of the past? We have

records of ice ages in sheets of clay with ice-borne boulders, of warm periods in mud from Finland containing the pollen of trees which do not now grow north of France. In dry countries the trees grow best in wet years, so every trunk is a record. When men begin to make records, we read of floods and droughts. Explorers tell when and where they met ice; armies crossed what are now sandy deserts. A great deal can be learned from the dates of harvests or those on which frozen rivers melted.

Twenty thousand years ago Britain and Canada were largely covered with ice, and no history was possible. Since then the climate has sometimes been better than now, sometimes worse. About 6,000 years ago it was much warmer on both sides of the Atlantic. Oysters grew on the coast of Eastern Canada, which is now much too cold for them. In Europe men were using polished stone tools and beginning agriculture.

Then came a period of drought over an area stretching at least from Britain to Siberia. The Irish bogs were overgrown with forests. A two-storey oak house of this period, built with stone tools, was found resting on 14 feet of bog, with 26 more feet of bog above it.

About 1800 B.C. the weather became cold and wet for about 2,000 years. Bogs drowned the Irish forests, and in England the river valleys were choked by marshes. But the chalk downs, which are now dry, were thickly populated, with brooks running in what are now waterless valleys. The Mediterranean region was much cooler than to-day, and Central Europe was so cold that armies often crossed the Rhine and Danube on the ice.

This was the state of affairs when the Romans

conquered Britain. The rainfall was so great that Roman wells have been found whose bottoms are 100 feet above the present water level. Yet we know they were used, because oaken buckets have been found in some of them.

About A.D. 450 the weather suddenly changed, and until A.D. 1200 was much warmer and drier on an average. This change had a great influence on English history. The Saxon invaders were able to cross the North Sea in much smaller ships than would be safe to-day. They probably found the lowlands largely unoccupied. And the Britons on the uplands were probably in the grip of an economic crisis due to the drought.

We are taught at school that the Saxon invasions were due to the withdrawal of the Romans and the treachery of the British King Vortigern, who invited the Saxons to help him. This may be true, but kings do not invite foreign armies to help them if their subjects are prosperous and contented. The great change in the weather was probably one of the straws which, so to say, broke the back of the dying economic system of the Roman Empire.

During the warm period the Vikings not only invaded England, but Iceland and Greenland; and some of them reached North America. Their records do not record ice as a danger till A.D. 1200. After that date their descendants were gradually frozen out of Greenland. Of course, the weather was not uniformly good. There were fifty years or so of heavy rainfall from A.D. 1000 to 1050.

Since A.D. 1200 the weather of Britain has had its ups and downs, but on the average it has been better than in Roman, and worse than in Saxon, times. We have recently had a series of mild winters, but not a long enough series to suggest that it will continue for centuries.

However, a weather change to-day would have little effect on Britain unless it were greater than any since the last ice age. Most of us work indoors. The introduction of field drains in the eighteenth century affected British farming more than most climatic changes. Irrigation is bringing fertility to areas of Soviet Asia which had been deserts for thousands of years. Whatever he may have been in the past, man to-day is no longer the slave of the weather.

STARS, INCLUDING THE EARTH

THE MOON ECLIPSED

NEXT Monday, November 7th,[1] there will be a total eclipse of the moon, and if the weather is fine, it should be easily seen from this country. The moonlight will begin to fade at 8.41 p.m. The eclipse will be total from 9.45 till 11.8, and will be over at 12.12 a.m. So look out between 10 and 11 p.m., even if you don't think the whole show worth watching.

It has long been known that lunar eclipses occur when the moon is full—that is to say, when the sun, earth and moon are roughly in a line, in that order. If the moon's orbit round the earth were exactly, or almost exactly, in the same plane as the earth's round the sun, then there would be an eclipse at every full moon. But if we imagine the earth's path round the sun as lying in a horizontal plane, say approximately 10 yards across, the moon's orbit would be the size of a halfpenny, slightly tilted.

We know the angle of tilt, and it is easy to calculate roughly when an eclipse will occur, though to do so within a minute, or even a few seconds, is another matter. To understand just what happens, it is easiest

[1] 1938.

to imagine ourselves on the moon, wearing a suit containing oxygen under pressure (for there is no air) and insulated against heat and cold with several layers of aluminium foil.

A man on the moon would describe what we call a "lunar eclipse" as an eclipse of the sun. When the moon is near the sun, and therefore "new," we cannot see it from the earth. But the earth is probably never invisible from the moon. For some of the sun's rays which pass through our air are bent from their straight paths, so a little light gets round the edge of the earth.

A man on the moon would first see part of the sun's disk covered by the earth. So less sunlight falls on the moon, and those parts of it where the sunlight is diminished are said to be in the earth's penumbra or partial shadow. After a while the sun is totally eclipsed on a portion of the moon, and we see the earth's shadow creeping across it.

But this shadow is not completely black. It is generally of a dull red colour. While the earth hides the sun from the moon, a certain amount of light gets round the earth's edge through our air. And this light is reddish, like the light which comes to us after the sun has set. So the man on the moon would first see a bright red patch on the earth's edge at the point where the sun has just disappeared. This would gradually extend to a red ring round the earth, and then concentrate into a red patch where the sun was due to appear again.

Naturally enough, the moon appears red to us when it is illuminated by this red light. All this, you may say, is theory. It looks all right on paper, but how can one check it? Quite easily. We can find out the particular spots on the earth which will be just in a line between the sun and moon when the total eclipse begins and ends. If there is a bright red sunset as the eclipse begins, the moon will get unusually red light, and, similarly, a bright red sunrise will mean a red end to the eclipse.

These observations have been made, and check up pretty well. Therefore the theory is a good theory. Whenever anyone puts forward a scientific theory, ask this simple question: Does it enable us to predict or control some event which we could not predict or control before? If so, it is a valuable theory, even if it is wrong. Because if we can definitely disprove a theory, we have at least got a bit of new knowledge.

So the theory that the moon is moved by a special angel is a bad theory, because it does not enable us to predict anything. But the theory that the moon moves round the earth in a perfect circle, as some astronomers believed 2,000 years ago, was a good theory, though false. It enabled predictions to be made which were nearly, but not quite right. And just because they were wrong, it was disproved. So it was a step towards the present theory, which is still not perfect, because the moon may be several seconds earlier or later than its calculated time.

Eclipses of the sun and moon have been recorded for over 4,000 years. These records enable us to do two things. In the first place, we can date some past events. For example, the earliest known date is 2283 B.C., when a total eclipse of the sun occurred at Ur, in Iraq, just before it was taken by the Elamites.

They also enable us to correct our calculations. For the earth's spin is very slowly decreasing, and the days getting longer, while the moon is moving away, and the length of the month increasing. So calculations based on modern figures are an hour or so out when we go 3,000 years back. In fact, not only natural objects, but the laws of Nature themselves, which are only human accounts of natural regularities, are changing with time.

So if you see the moon being eclipsed, you may congratulate yourself that we (or at least astronomers) know a good deal more about it than our ancestors did. But we still have plenty to find out. For example, how quickly does the moon's temperature fall? Does the red colour agree exactly with our present theories, or only roughly? Can we use its spectrum to check the theory which has led to the analysis of the air on other planets? (See p. 66.) And so on. Though the cause of lunar eclipses has been known for over 2,000 years, science goes on, and we have still plenty to learn about them.

THE SUN ECLIPSED

On Wednesday, April 19th,[1] the sun will be eclipsed. In London the eclipse will begin at 6.30 p.m. (summer time) and about a third of the sun will be covered at 7.15. As the sun will be low in the sky, it should be quite safe to look at it without darkened glasses.

A little more of the sun will be hidden from Scotland

[1] 1939.

than from England, but nowhere in Britain will half the sun's disk be dark. In many cases, when the sun is eclipsed we read of expeditions going to some remote island or desert to observe it. But this time there will be no such expeditions, because the eclipse will nowhere be total.

An eclipse of the sun occurs, of course, when the new moon comes exactly between the sun and the earth. Sometimes the eclipse is total in some parts of the earth—that is to say, the moon hides the whole of the sun. But on April 19th observers in Alaska and perhaps the crews of a few ships in the Arctic Ocean will see an annular eclipse. That is to say, the moon will not cover the whole sun, but where the eclipse is deepest a ring of sun will be visible round the moon.

It is a rather remarkable coincidence that the sun and moon have almost exactly the same apparent size, though the moon usually covers a little more of the sky. If the sun and moon were always at just their average distances from the earth, then all eclipses would be total, provided a line through the centres of the sun and moon reached the earth somewhere.

But actually the earth moves round the sun, and the moon round the earth, in orbits which are not quite circular, nor indeed quite elliptical. When the sun is as far away as possible and the moon as near as possible, a total eclipse may last as long as seven minutes. When the sun is near and the moon far, as next week, the eclipse is annular.

Nowhere else in the solar system does this happen. The moons of the planet Mars are too small ever to hide the sun. Some of those which go round Jupiter always cause total eclipses on that planet, for their shadows can be seen crossing its face, as an astronomer on Venus could see the shadow on the earth caused by a total eclipse.

Still odder is the fact that there was a first annular eclipse of the sun. The average distance of the moon from the earth is very gradually increasing, and when it was nearer it could always hide the sun completely. In the same way, there will be a last total eclipse. But the first annular eclipse happened a long time before there were any men to see it.

And the last total eclipse will be many million years hence, when our descendants will probably be very unlike ourselves, and capitalism will be a tradition of the very remote past, perhaps something of which only historical specialists ever learn. The date can be roughly calculated, but the calculation would take me several hours, and I am busy with A.R.P.

It is lucky that man evolved before the last total eclipse, for total eclipses have helped knowledge and assisted progress. They were very alarming events, so men tried to deal with them in various ways, such as beating gongs to scare the dragon which was trying to eat the sun. And when it was possible to predict them, this meant a great increase in human self-confidence.

We are still mostly in the gong-beating stage as regards politics and economics. It is not generally thought that history is a science and prediction possible. On the contrary, Marxists who try to apply scientific method to history are treated much as people were treated a few thousand years ago who said that the black circle creeping over the sun was not a dragon or a devil, but only the moon.

The success of the scientific method in predicting eclipses was a great step forward. As soon as the theory was so accurate that an expedition could be sent to a spot where a total eclipse was due, a lot more information became available. When the sun's disk was covered, solar prominences—that is to say, luminous clouds a long way above the edge of the sun—were seen, and also the corona, a glow stretching out still further.

More recently Einstein predicted that the light from stars passing close to the sun would be bent out of its path by gravitation. The idea was not new; indeed, the French revolutionary leader, Marat, had thought, on quite insufficient evidence, that light was deflected by gravitation. However, Einstein's prediction was accurate, not merely as to the bending of the light, but as to its magnitude.

But the observations are not yet accurate enough to make it certain that Einstein was completely right. Further work may show that his theory will have to be slightly modified. Other recent developments include photography of the corona from high-flying aeroplanes, which enable details to be seen which are invisible from the earth's surface, and observations on the effect of eclipses on radio transmission.

In fact, there is so much to be learned from total eclipses that we may be glad that annular eclipses are rather uncommon. But at least we can predict them. So no fruitless expeditions will be sent to Alaska. Whereas we cannot yet predict the weather except for a few days ahead. Hence many eclipse expeditions are spoiled by cloud, and very likely those who look for a partial eclipse as a result of this article will be disappointed.[1]

IS THERE LIFE ON THE PLANETS?

DURING this month[2] all the five planets which can be seen with the naked eye are visible. Venus is an evening star, setting about an hour and a half after the sun. Jupiter is the brightest star in the sky after Venus has set, and Saturn is to be seen to the east of Jupiter. If you go to work in the morning before daybreak, you may see Mercury and Mars in the east, rising before the sun.

What are these planets? Many primitive people thought they were gods and controlled human destiny. Their reasons seem to have been something like this. The fertilising flood of the Nile comes just after Sirius is first visible in the morning. So Sirius controls the Nile flood. And in the same way another star controls the lambing season and a third the wheat harvest.

But some events, such as wars and pestilences, do not happen regularly. So they must be controlled by the stars which move their position relative to the others, the planets or wanderers. This argument is no worse than many which we now hear about the causes of wars and

[1] They were not, at least, in London.

[2] September, 1938.

slumps. Even 2,000 years ago, though the Greeks and Romans did not usually worship the stars, they thought it blasphemous to suggest that heavenly bodies were made of the same sort of stuff as earthly things.

It had long been clear that the moon shines by reflected sunlight. When Galileo turned his telescope on Venus, he saw a crescent like the moon's, which altered its shape as the planet moved. It was therefore clear that the planets are cool bodies like the earth. Copernicus' theory that the planets and the earth went round the sun made it possible to calculate their distances, and their sizes were then determined by measuring their images in a telescope. Venus and Mars turned out to be about as big as the earth, Mercury somewhat smaller and Jupiter and Saturn much bigger.

It seemed natural to speculate that they were inhabited. But before it was possible to say whether life as we know it can exist on the planets, a lot more information was needed. And our knowledge of the planets has not increased very greatly in the last fifty years, although we have found out vastly more about the distant stars and nebulæ.

The reason is interesting. If we want to know more about a distant cluster of stars, we train a telescope on it, and use very complicated machinery and a still finer human control, so that the telescope follows the cluster in its apparent motion across the sky. We then take a time exposure lasting for many hours.

But we cannot photograph the surface of Mars in this way, because the planet turns on its axis about as quickly as the earth. So astronomers must rely on their eyes. And as a matter of fact, some of the best observation of the planets is done by amateur astronomers, including Mr. Will Hay, the comedian, and several English country clergymen, using relatively small telescopes.

It is clear that we can see the solid surface of Mars, whereas in the case of Jupiter, and probably Venus and Saturn, we can only observe the tops of clouds, which may consist of drops of liquid or of solid dust. We can follow seasonal changes on Mars. During the winter each pole develops a white cap, which doubtless consists of frost. This frost may be frozen water. But it may be solid carbon dioxide, which is used in the refrigerating industry under the name of "dry ice," and is only solid at temperatures far below the freezing-point of water. There are also colour changes elsewhere which may be due to vegetation.

Although we have not learned much fresh about the surface markings of these planets in the last fifty years, they have been studied with two instruments which tell us a lot about them. A sensitive thermopile placed at the focus of a telescope gives an electric current proportional to the heat coming from a star. Reflected sunlight is not stopped by a thin layer of water. But heat from a body which is warm, but not white hot or even red hot, is stopped. So by putting a little water bath in front of the thermopile we can measure the temperature of the planets.

Mercury and Venus are hotter than the earth, but Venus is probably well below the boiling-point of water. On the other hand, Mars is colder, though at least in the daytime in its tropics ice would melt. But the visible surfaces of

Jupiter and Saturn are intensely cold, though the solid surface under the clouds may be somewhat warmer, especially if there are volcanoes.

We can also use the spectroscope. When light is passed through carbon dioxide, certain components of it are absorbed. Not visible light (or carbon dioxide would be coloured), but infra-red light, which can be photographed on a specially sensitised plate. The same is true for other gases. So by comparing sunlight reflected by Venus with sunlight reflected from the moon, we can see that the former has passed through the equivalent of several hundred yards of pure carbon dioxide at the ordinary pressure.

And there is no oxygen or water vapour in the atmospheres of Venus or Mars, or at any rate far less than on earth. Hence a man could only live on these planets in something like a mine rescue apparatus, and it seems to me a little unlikely that there is life of any sort on Venus. If there is life on Mars, it is probably more like that of the bacteria which live without oxygen in black mud than to those of familiar animals and plants. So perhaps we had better make our own planet fit for rational beings before we colonise others.

SHOOTING STARS

Now that the nights are getting longer we have more chance to look at the stars, and when we do so we sometimes see a bright point shoot across the sky and fade out.

In the past some people thought that these were stars which had fallen from their places. But we now know that stars do not disappear, and, moreover, that most of them are much bigger than the whole earth.

Mohammed thought that shooting stars were thrown by the angels at devils who were trying to overhear the secrets of heaven. Only recently have we learned much about these objects, which are the only things large enough to be visible which come to our planet from outside.

If two people some miles distant record the flight of the same meteor they see it against a different background of stars. If two photographs are taken, things are still better. The tracks generally appear farther apart at their ends than at their beginnings, because the nearer the meteor to the ground the greater the apparent distance will be.

In this way the speed and height of meteors can be measured, and between 1931 and 1933 a team of six observers measured them for many thousands of meteors at two stations twenty-two miles apart in Arizona, where nights are very clear. They looked at the sky through a wire network, and noted each track and its time on a sheet of specially ruled paper. But that was the least part of the job. Opik, an Esthonian astronomer, has been calculating the results ever since, and hasn't finished yet.

Still, the average height turns out to be about fifty-five miles, and the average speed about thirty or forty miles per second.

In fact, there is no doubt that meteors are simply hard bodies moving so fast that they are made white-hot and finally explode into flaming vapour. How big are they, what are they made of, and where do they come from?

A few are so big that parts of them fall to the ground, and thousands have been collected. A very few are so large as to cause widespread disasters. One such fell in a Siberian forest in June, 1908, making a huge hole, such a wind that it blew down trees for many miles around, and a sound which finally became a pressure wave recorded by sensitive barometers in England. If it had fallen in London or even in the Thames Estuary it would have killed some millions of people.

These large meteorites consist of an alloy of iron and nickel, of stone rather like certain volcanic rocks, or generally of a mixture of these. Paneth, a German professor now a refugee in England, determined their age from the helium produced in them by radio-activity, and found that they had been solid, and presumably flying through empty space, for times varying from a few million up to 2,000 million years.

However, only about five meteorites are observed to hit the ground every year, and few or none have been so carefully observed that their speed is known so that we could tell where they came from. This is known for some of the smaller shooting stars which arrive in swarms at certain times of the year, each swarm moving parallel, so that to an observer on the ground they seem to scatter from a particular point in the sky.

Calculations show that they must move round the sun in elliptical paths like those of comets. Actually some of them move in the same paths as comets. Thus the well-known meteors which appear to shoot from the constellation of the Lion in November, move in the same path as Tempel's comet, seen in 1866. The comets seem to consist of dense swarms of meteors, while others are scattered along their orbit.

But, of course, comets are quite different from meteors. The comets are huge masses of dust and gas, generally millions of miles away, and shining by reflected sunlight, while meteors shine because they are hot. Their trails left in the air are only fifty miles or so above ground, and disappear in a few minutes. So the news paragraph which I regret to say recently appeared in the *Daily Worker* as well as the capitalist newspapers, describing a comet seen over France only, was nonsense.

The comets and meteor swarms rarely if ever travel so fast as to escape from the solar system. They are parts of it, and were very likely produced by volcanic eruptions on Jupiter or other planets. But quite a number of the shooting stars which appear one at a time are moving at speeds so high as to show that they came from the space between the "fixed" stars.

For a meteor going at a speed more than 41 per cent. above the earth's relative to the sun, will fly off again into outer space unless it hits something; and must have recently come into the solar system from outer space. The observed speeds also show the effect of the earth's motion through space. The fastest meteors relative to the earth, which flare up at the greatest height, come from the direction in which the earth is moving in its orbit, which is roughly the part of the sky overhead at daybreak.

In fact, shooting stars behave as you would expect them

to on common-sense grounds. That is why, in a country which is running away from common sense, you read so much about paradoxical particles such as cosmic rays and so little about shooting stars.

WHY EARTHQUAKES?

OUR ancestors used to think of the hills as everlasting. And society discouraged them from even thinking about change. The king was on top, the nobles and priests below him and the workers at the bottom. This was supposed to be the divine plan of society, and Nature also was made according to an unchangeable plan. If the solid earth was moved, this was due to divine intervention, and it was impious to enquire too closely into such movements.

We now know that neither the form of the landscape nor of society is fixed. Mountain ranges may sink under the sea and autocratic empires become people's republics. The laws which govern these changes can be discovered and these discoveries can be used.

After a big earthquake it is often found that there has been movement along a fault. A fault is a crack passing through the earth and the rocks below it, and one side of the crack may move up, down, or sideways, relative to the other. Faulting is familiar to coal-miners. A coal seam suddenly comes to an end in hard rock, and to find it again it may be necessary to go many yards up or down. These faults were made by earthquakes millions of years ago, when the British hills were being raised out of the sea.

But in California mountains are still being made, and at most earthquakes one or other of the faults which already exist acts as a zone of weakness. This does not deter speculators from building houses along the faults. And, of course, legislation to prevent this would be an interference with private enterprise, unthinkable in a state which imprisoned Tom Mooney for twenty years on false evidence.

But a study of zones of weakness in the earth's surface tells us nothing of the forces which make them give way. Earthquakes are like revolutions. They are the product of strains which have been piling up for a long time. What is the ultimate cause of these strains?

Probably the main cause is the cooling of the earth. Its inside is still very hot, as every miner knows. The heat escapes, though very slowly, and the earth contracts, at a rate of a few inches a century at most. Now, the inside of the earth is not exactly liquid, but it gives way fairly easily under a prolonged strain. The outer fifty miles or so are floating on the molten depths. This is shown by the fact that mountains are built of lighter material than plains, and the heaviest rocks are found under the bed of the ocean. If mountains were made of heavy rock they would soon sink down.

The almost liquid interior contracts, and the hard crust has to give way, and folds into mountains and ocean troughs. At present the most important folding is going on along the shores of the Pacific Ocean, where great mountain chains are close to the coast, and only a few miles out are troughs deeper than anything in the Atlantic ocean. There are a few rifts in the earth's crust, where it has pulled apart, but these are very much rarer than the folds.[1]

In the folded regions most earthquakes are not directly due to the folding process, but to the strains caused by the folding, which are very gradually relieved. Another kind of earthquake is caused by the action of water and ice in wearing down mountains. Huge quantities of mud, sand and gravel are washed down from the Himalayas on to the plains of India and laid down in the sea at the mouths of the Ganges and Indus. This causes an extra loading on the rocks below the Indian plain, and occasionally they give way, causing an earthquake.

Earthquakes are, of course, common near volcanoes, and it used to be thought that volcanoes caused earthquakes. The truth is rather the other way. Where the rocks are folded and cracked, molten matter and steam from below can readily find their way up. So some, but not all folds, are studded with volcanoes.

The modern study of earthquakes was founded by an Englishman, Milne, who went out to Japan to teach physics, and constructed instruments to record the waves produced by earthquakes at a distance. Nowadays these instruments are found in many different countries and have, of course, been greatly improved.

When an earthquake occurs many miles away, it starts several sets of waves going. Those which travel quickest through the rocks—at about three and a half miles per

[1] This is the view of most geologists. For another opinion, see page 74.

second—are called push waves, and consist of a backwards and forwards movement too slight to be felt, but big enough to move a specially designed instrument. Sideways movements, called shake waves, travel a little more slowly.

So the further away the earthquake is, the bigger the interval between the push and the shake, and in this way one can calculate how far the waves have travelled, and therefore the distance between the observatory and the earthquake. So from the records of three observatories the exact site of the earthquake can be determined. In the same way it can be discovered at what depth below the surface the shock started.

So far we have discovered a good deal about the strains which cause earthquakes, but not much about why the rocks snap on one day rather than another. They are rather commoner when the moon is full or new. The same force which causes high tides strains the earth, and may be the last straw, so to speak, which makes the rocks give way.

Some years ago a Japanese biologist got together evidence that certain fish were disturbed for some hours before an earthquake. If this is true they could be used, like watchdogs, to give warning. But at present Japanese scientists are doing more to destroy Chinese cities than to save their own. And if the British Government spent as much on medical research as it does on military research, including the design of bombing aeroplanes, we could justifiably look down on the Japanese as barbarians.

At present the best defence against earthquakes is

provided by earthquake-proof buildings. But in days to come, even if we cannot prevent earthquakes, we may be able to predict them so accurately that by evacuating the area where one is expected, loss of life will be wholly prevented.

DO CONTINENTS MOVE?

So far in this series of articles I have dealt with matters on which the majority of scientific workers are agreed. To-day I describe a theory which is held by a minority of geologists, and on which I am not myself qualified to judge. Many of the scientific controversies of to-day are hard to explain in a short article. This one is easy, though, of course, the evidence on each side cannot be given.

Twenty years ago almost all geologists thought that, although the continents had moved up and down, and some mountains had been formed by folding, they had stayed in the same place, or nearly so, relative to one another. If London is 3,200 miles from New York to-day, this was assumed to have been so when the rocks beneath the cities were formed.

The first men to doubt this in a systematic way were two American geologists, F. B. Taylor and H. B. Baker, in 1910 and 1911; but Wegener of Graz, who was killed while exploring Greenland in 1930, carried these ideas a good deal further, and now many geologists think that the continents have moved relative to one another and are still moving.

A glance at a map of the world shows that if America were moved eastward in a block, the Atlantic Ocean would be filled up, and the British Isles would be brought close to Newfoundland and New England, while West Africa fitted against the West Indies, Brazil into the Gulf of Guinea, and so on. Baker believes that they fitted together in this way in the past.

The rocks, especially the older rocks, at corresponding points on opposite sides of the Atlantic correspond very well, or so it is claimed. Thus the coal-fields of South Wales appear again in Pennsylvania, the diamond-bearing beds on the coast of Brazil correspond to those of South-West Africa.

If the same jigsaw theory is applied in the southern hemisphere, Australia, Antarctica and India can all be fitted on to Africa. If so some very strange facts are explained. At the time when the coal was being formed in England, tillite—that is to say, boulder clay now converted into soft rock—was being laid down in South America, South Africa, India and Australia.

Clay containing boulders is only produced by ice, and the theory holds that all these continents were then grouped round the South Pole, while coal measures were being formed in the tropics. If these continents were once united, a number of very queer facts about the geographical distribution of animals are explained. For example, marsupials, like the kangaroo and opossum, are mainly found in Australia and New Guinea, but also in South America; *Peripatus*, a strange animal like a caterpillar which never becomes an insect, is found in New Zealand, Australia, South Africa and South America.

In their drift apart, these southern continents have made the Indian and South Atlantic Oceans, while the Pacific Ocean represents the remains of a much larger original ocean, against which the Americas are pressing, piling up mountain ranges as they move westwards. Wegener thought that all the continents once formed a single mass. Du Toit and others think that the southern continents have approached Europe and Asia. The pressure of Africa against Europe made the Alps, Balkans, and Atlas Mountains, whilst the northern part of India actually penetrated under what is now central Asia, and lifted up the huge plateau of Tibet.

Africa is still splitting to-day, along the rift valley in Kenya, and in other places, and it can be predicted that in the course of millions of years much of East Africa will split off from the main continent, as Arabia and Madagascar have already done. Determinations of the longitude of places in Greenland seem to show that it is moving away from Europe at 20 or 30 yards per year. If this is finally proved to be so, there is no doubt that the theory of drifting continents will have to be accepted.

But how can continents move like this? We now know that they are made of rocks which are a good deal lighter than those composing the ocean floors, and can be regarded as floating on them. The heavy rocks are thought to be softer, and to crumple or stretch as the continents move. But it is hard to imagine what can be the huge forces which move the continents.

Perhaps the most hopeful theory is that of Holmes,

which attributes it to currents of molten rock under the continents, rising in the hot centre of the continents, moving outwards, and sinking where the rocks have been cooled down by the ocean. Many geologists dismiss these ideas as absurd, but at present a number are supporting the new theories, and what is to-day a heresy may be orthodox to-morrow.

Among the strongest adherents are some economic geologists, such as the American Association of Petroleum Geologists. When we remember that geology owes so much to Smith, an engineer who classified the English strata when supervising the excavation of canals over a century ago, and to those who carried out geological surveys primarily for economic purposes, we may be inclined to back the practical men against the theorists who say that their views are impossible.

AMMONIA

THE explosion in the Eldorado ice-cream works,[1] where one girl was killed and many injured by ammonia, has brought this substance into the news. What is ammonia? If you buy a bottle labelled "ammonia" at the chemists, you do not get a pure substance, but a solution of ammonia in water.

Pure ammonia is, at ordinary temperatures and pressures, a gas. But it can be liquefied, not only by cold, like

[1] In July, 1938.

oxygen or nitrogen, but also by pressure. It needs about 7 atmospheres, or 100 lb. per square inch, to make it into liquid at ordinary temperatures, and if this pressure is taken off, it immediately becomes a gas again.

Now, when a liquid becomes a gas, a lot of energy is needed before the molecules which have been confined in the liquid can fly out freely into the air. This energy is provided by heat. Either the liquid must be heated, as when water is boiled, or it takes away some of the heat from the liquid and makes it cooler.

This is the principle of an ammonia refrigerator. The gas is compressed into liquid, which heats it up. It is then cooled down to the surrounding temperature. When it is now allowed to expand, it becomes very cold, and will freeze water.

Ammonia is rather a rare substance in Nature, unless we go a long way off. On our earth it is produced in small amounts by decaying urine and other substances. But on the planets Jupiter and Saturn it is extremely common. This fact was only discovered in the last ten years by means of the spectroscope.

When you look at white light through a prism and focus it properly, it is broken up into its constituent colours. Each kind of atom or molecule absorbs a special set of vibrations, its signature tune, so to say. For example, mercury vapour, when hot or electrically excited, gives a very characteristic green light which is familiar in advertisements and modern street lamps, and if a strong light shines through hot mercury vapour the same green rays are absorbed.

Now, ammonia does not consist of single atoms, but of

1 atom of nitrogen united into a molecule with 3 of hydrogen. Molecules have a much more complicated spectrum—that is to say, set of rays—than atoms. And ammonia gas is nearly transparent. It was not till light had been passed through many feet of the compressed gas that the spectrum was discovered.

It was then found to be the same as that observed in the atmospheres of the planets Jupiter, Saturn and Uranus. That is to say, the sunlight passes through the "air" on these planets, and is reflected from their clouds. And on the way it loses the rays characteristic of ammonia.

Instead of a mixture of nitrogen and oxygen, as on our earth, the outer planets have atmospheres of hydrogen, ammonia, and methane, or fire-damp. These planets are so cold that water would be permanently frozen. But they may have oceans of liquid ammonia, methane, or other substances which are gases on our earth.

Life as we know it would be impossible. But just as many ordinary chemical reactions take place in solution in water, so others occur in liquid ammonia under pressure. In fact, the American chemist, Franklin, has worked out the beginnings of a system of chemistry in which ammonia takes the place of water. So if there is any life on the outer planets, it is probably based on chemical reactions of this kind.

However, for almost all the living creatures on our earth, ammonia is poisonous, as the unfortunate girls at the Eldorado works found. Even its compounds, such as ammonium chloride, are deadly if injected into the blood, causing violent convulsions.

But ammonia is formed during digestion and other vital processes, and has to be immediately got rid of, or we should all be poisoned in a few hours. So it is converted by the liver into a harmless substance called urea. (I have eaten a $\frac{1}{4}$ lb. of it at a time, so I ought to know that it is not a poison.) This is then got rid of by the kidneys.

I took advantage of this fact to make myself acid. I wanted to reproduce in myself the symptoms of acid poisoning, such as often occur in people dying of diabetes or kidney disease. So I drank as much hydrochloric acid (spirits of salts) as I could. But I couldn't drink enough without burning my stomach.

So I drank a solution of ammonium chloride, the compound of ammonia with hydrochloric acid, trusting my liver to turn the ammonia into urea, which it did. The acid was left behind, and I was able to produce the symptoms of acid poisoning, such as shortness of breath and several more complicated ones.

This showed that acid poisoning was rather less important in killing people than had been thought, and also provided a simple means of making people more acid when desired—for example, for dissolving radium out of the bones of girls with radium poisoning. Experiments of this kind are safe enough if one treats oneself as a chemical system, and makes no mistakes either in the calculations or the preparation of the chemicals to be eaten. If one makes mistakes of this sort, the world contains one bad biochemist less!

WHAT PAPANIN'S POLAR EXPEDITION FOUND

THE first scientific results of the Soviet polar expedition are now being published, though the complete account will probably not be available this year. But we can already state something of what they found. Previous expeditions had got to the Pole or near it, but they had been too busy getting there and back to make many discoveries on the way.

One of the most important tasks of the floating camp was to take observations of the sun or stars whenever weather permitted, so as to find their own exact position in the Arctic Ocean. Thus the drift of the ice could be ascertained. It was, of course, a good deal quicker than had been expected. They had hoped to drift for a year, but had to be taken off after 274 days.

But they did much more than determine the drift. They discovered the reason for it. The ice is moved by the wind. Its drift relative to the main body of water below is about one-hundredth of the wind's speed. Owing to the earth's rotation, the ice deviates to the right of the wind's direction. This could only be discovered by dropping cables with a special type of current-meter which measures the drift of the ice relative to the water at different depths.

The general movement of the surface water down to a depth of about 700 feet is determined by the prevailing winds, which on the average blow towards the Atlantic Ocean. This information will enable the Soviet meteorologists to predict the movements of ice in other parts of the Arctic Ocean from a knowledge of the winds; and this will be of great value for the navigation of the northern coast of the Soviet Union. It should help to prevent ships from being caught and crushed in the ice in future, as the *Chelyuskin* was in 1934.

At a depth of 700 to 2,500 feet the surface current of cold water towards the Atlantic is balanced by a warmer but salter current from the Atlantic towards the Pole. It will be necessary to observe these currents in a number of different years. When this has been done they may be found to play a big part in determining the temperature of the surface water of the North Atlantic Ocean, and hence the weather of western Europe.

Besides the observations of weather and drift, magnetic observations were made every day by Feodorov. Of course, the compass does not point to the North Pole, but roughly towards the North Magnetic Pole in northern Canada. The deviation of the compass from due north is very exactly known in most parts of the world, and it could be predicted in the Arctic Ocean. However, these predictions turned out to be a little bit out in some places.

The exact determinations which were made until the floe began to spin too rapidly will be of great use to future air navigators (the planners of the Soviet Union look forward to a regular air route across the pole to America) and perhaps even of some value to future expeditions on the ice surface.

A number of soundings of the ocean depths were taken. The Arctic Ocean turns out to be over 14,000 feet deep in some places, though one submarine mountain range 4,000 feet high was discovered. These observations came to an end because the winch for winding up the cable was naturally placed at the edge of the ice floe, and drifted away from the camp when the floe broke in two.

Among the most interesting results of the expedition were two samples of mud from the bottom. The surface layer was reddish brown, but below this the mud was grey. When these samples have been examined with the microscope and chemically analysed they may tell the first chapter of a very big story.

The bottom of the oceans is covered with mud of various kinds, the commonest sorts being a greyish chalky ooze consisting of microscopic shells such as make up the chalk downs of England, a siliceous kind consisting of microscopic plant remains, and a red clay which is thought to consist of volcanic ash or meteoric dust, the remains of shooting stars.

Only last year the American scientists Piggot and Mackey invented a gun which is lowered to the bottom and shoots a brass tube 10 feet long and 1 inch or so in diameter into the ocean floor. The tube is pulled up with a sample of the upper 10 feet of mud. Where the mud is not being laid down too quickly these tubes told a clear and simple story. The top layer in the North Atlantic is mainly made of the shells of animals now living in the surface water.

But below that there are two layers full of volcanic ash, perhaps from a colossal prehistoric eruption in Iceland, and four layers of shells of microscopic animals now living in the Arctic. It was known from a very careful analysis of the clay and stones left behind by glaciers which once covered all Switzerland and some of Germany that there have been four ice ages in the last million years or so.

But nowhere on land has such a continuous record been laid down as on the ocean bed. Will the Arctic Ocean bed tell of a warmer time when the polar sea was not covered with ice? The British climate was much warmer about 6,000 years ago, and there may then have been no ice at the pole during the summer. Or the change may be the record of a volcanic outburst. Probably the next Soviet polar expedition will take a Piggot gun and settle the matter.

Shirshov, the biologist, found plenty of life below the ice. Nansen thought the Arctic Ocean was a desert. But Shirshov found that when the sun shone through the ice microscopic green plants began to grow, as they do in the seas round England every spring. And, just as here, tiny water-fleas feed on the plants.

In the North Sea these water-fleas act as food for herrings and other fish. I do not know whether Shirshov caught any fish, but it is fairly sure that there must be fish below the ice. And it is even possible that in future there will be drifting fishing stations on the Arctic ice.

Besides this the force of gravitation was measured. The greater the force the quicker a pendulum swings. So if the rocks underground are dense it swings very slightly faster than if they are light. Up till now the only observations of gravity at sea had been made in submarines. Those made on the ice are probably more accurate. They show that the rocks under the ocean bed are on the whole a good deal denser than those which make the continents.

The first explorers of the Arctic Ocean were Englishmen, such as Sir Hugh Willoughby, sent out by the Company of Merchant Adventurers in 1553. For nearly three centuries after this Britain led the world in exploration, and such great men as Drake, Cook and Livingstone played an important part in mapping the world. With the coming of monopoly capitalism this is no longer so. Effort is more and more concentrated on extracting big profits from existing resources rather than on discovering new ones. And it is natural enough that the lead in exploration has passed to a people who do not restrict production, but can find a use for all that they discover.

MATHEMATICS AND PHYSICS

PROBABILITY—AND A.R.P.

ON few branches of science is the public more completely misled than on the theory of probability. It is easy to understand why. Gambling plays a very important part in our social life, whether in the form of football pools for the poor, or Stock Exchange speculation for the rich. The people who make big fortunes out of gambling do not want the rest of us to think too much about it. They would rather we bought "lucky charms" than grappled with the necessary calculations.

One of the hardest things in the whole theory is to understand how a number of chances combine to make a certainty. "One knows," wrote Engels in *Feuerbach*, "that what is maintained to be necessary is composed of pure accidents." Of course, it takes a great many independent pure chances to make a practical certainty. We had better not take tosses of a coin as an example. For a coin may be biased for heads or tails. It would be safer to look at the numbers of passing cars, and call an odd a head and an even a tail.

If we count 1,000 cars we are unlikely to get just 500 heads. But the chances are just over even that we shall get between 510 and 490—that is to say, right within 2 per cent. And if we looked at a million cars, the odds against being 1 per cent., or 5,000 out from the half million, are a number of twenty-four figures to one. This is near enough to certainty for ordinary people.

What has all this got to do with A.R.P.? A lot. If we are thinking in terms of a future war with Germany, tens, and perhaps hundreds, of thousands of bombs will be dropped on Britain. The number of people killed by any particular bomb is a matter of luck. The number killed by 10,000 is nearer a certainty, if we had the facts on which to calculate it.

And we must use the kind of thought in planning A.R.P. that an insurance company manager or a bookmaker uses, and not the kind that a gambler uses. For example, if you are earning 25*s*. a week and want a motor car you will have to wait a very long time before you can save the price of one. But you may get one next week if you make the money in a football pool.

Some people think this sort of thing worth while, and they are impressed by the official arguments for dispersal during air raids—that is to say, the policy that people in a town which is likely to be bombed should not crowd together in one place where a single bomb might kill 100, but should scatter about, so that most bombs would kill a few, but no single one could kill very many.

This would perhaps be sound if only one bomb were going to be dropped on one town. It may be that it is better to have two people killed for certain than one chance in 100 of 200 people being killed. But if enough bombs are dropped, thousands of people will be killed, anyway, so the only thing that matters is to reduce the chance of being killed for each person. Provided everyone has time to get to a shelter, we can work out the chance of a person being killed if a bomb of a given size falls in the neighbourhood.

Of course, if he is in a deep enough tunnel, this chance is zero. But for other shelters we can calculate it roughly. First, suppose he is rich enough to own a garden and £8, which is what the Government says is needed for materials for timbering and roofing a trench, and that he has put in the necessary work and made a trench[1] 16 feet long by 4 feet wide. In order that a 50-lb. bomb should kill him, we may suppose that it must fall within the trench or 2 feet of it—say, an area of 160 square feet.

Another man is in a public trench in a park, 70 feet long by 6 feet wide. The same bomb will kill him if it falls in the trench or within 2 feet of the section 10 feet on each side of him—that is to say, an area of 500 square feet. So he is about three times as likely to be killed as the man in the garden trench. No doubt that is why the trenches in parks are almost always made long and perfectly straight instead of with many corners, as in the Great War and in Spain. Of course, such figures as these are very rough. But they give an idea of the way such chances can be calculated.

Again, consider the basements under modern steel frame and concrete buildings. They are not likely to be knocked down by the blast of bombs in the street. But if a heavy bomb actually falls through six cement floors into the basement, and bursts there, as has often happened in

[1] This was written before the time of Anderson shelters. The general argument is unaffected.

Spain, almost everyone in it will be killed. So your chance of being killed in a raid is proportional to the area of the basement, unless this is very large or divided up by stout walls.

I had originally meant to choose mouse-breeding as an example to illustrate the theory of probability. As our Government refused to agree to prohibit air bombing in 1932 or to protect us from it in 1938, it is an unfortunate fact that bombs are more suitable as instances.

SOME MATHEMATICAL CONUNDRUMS

Most normal children are thoroughly bored by mathematics at school, and no wonder, considering how they are taught. One reason for their boredom is that they are given hopelessly artificial problems to solve, instead of problems which arise from their daily life. Another reason is that mathematics are taught as something dead and finished, not as something alive and growing.

A child has some idea of what goes on in a chemical research laboratory. Men and women are trying to make new compounds, to measure the properties of compounds already known, or the speed of chemical changes, and so on. But what do professors of mathematics do when they are not teaching?

Most of their problems are too complicated to explain in a small article. But a few are quite simple. In science we are often satisfied if we find that something is true in a very great many cases, even if we cannot prove that it must be true. For example all leguminous plants (e.g. peas and clover) so far observed have two cotyledons or seed leaves, but we cannot show that this must be so.

But in mathematics this is not good enough. The Babylonians knew that the angle in a semicircle is a right angle, that is to say if you join a point on a semicircle to the two ends, the lines make a right-angle. But it was a Greek called Thales who first showed that this must necessarily be true. And the whole idea of proving mathematical theorems started from this simple beginning.

There are quite a number of facts in mathematics which are known to be true; but no one has been able to prove they must be true. The simplest of all is the following. If a map is divided up into areas each of which is closed so that one can move all over it without crossing the boundary, then only four colours are needed to colour each area so that no areas of the same colour touch. Of course, five or more may meet in a point, but two of the same colour never meet in a line.

You can verify the truth of this on the counties of England. But can you prove that it is true for every possible map? If so I shall be delighted to propose you for a fellowship of the Royal Society next year! Or can you produce a map which needs five different colours? If so you will probably be able to get a fairly well-paid post as a mathematician.

What does it matter, you may ask? I have no doubt that if the four-colour theorem could be proved we should be able to make big advances in the theory of how molecules are arranged in solids and liquids, and this would lead within a few years to improvements in the practices of lubrication, polishing, and the like.

Here is another theorem which mathematicians have been trying to prove since 1742. Every odd number over 7 can be expressed as the sum of three odd prime numbers (that is to say, numbers such as 3, 5, 7, and 11 with no factors over 1). Thus 29 can be written as 23+3+3 or 19+5+5, or in several other ways. It has just been shown to be true for all odd numbers greater than some at present unknown number by a Moscow mathematician called Vinogradoff. And some day it will probably be proved completely.[1]

Its practical importance is this. Prime numbers occur with an irregular sort of rhythm. For example, in order to find them we take a table of numbers and first scratch out every second number over 2, then every third over 3, and so on. The primes are those which remain. And one of the mathematical methods used to study prime numbers is similar to that used for disentangling the irregular rhythm of a spoken telephone message into the regular rhythms of different musical notes.

So Vinogradoff's work will probably help those mathematicians who are working out methods for telephoning from Moscow to Vladivostok, or for sending 16 messages each way over a single telegraph wire between them. Though, of course, it will take several years before it finds a practical application.

Mathematics still suffer from the fact that the Greeks, who did so much to found it, belonged to a community based on slavery. So their intellectuals thought that

[1] I understand that van der Corput has since done so in Holland.

manual work was undignified, and never made machines to help them in their mathematics. And ever since then many mathematicians have despised mechanical aid.

But machines are now being made which can do in a day calculations which would take a lifetime if done on paper. One of these, invented by Bush of the Massachusetts Institute of Technology, is being used by Hartree in Manchester to solve problems about atoms and about machines for automatic control of furnace temperatures which were insoluble on paper.

But the academic tradition in most countries gives the highest place to "pure" mathematics—that is to say, mathematics with no obvious use—and looks down on the use of machines. When it is realised that these prejudices are products of the class struggle, mathematics will go forward more rapidly.

SEEING THE INVISIBLE

MAN is generally said to have five senses: sight, hearing, smell, taste and touch. In reality, we have several more. There are the senses of heat and cold, the sense of balance, by which we can distinguish up from down, the sense of muscular movement, and so on. Some animals have senses which we lack. Fish perceive water currents by special organs in tubes buried in their skin; and many people think that dogs and homing pigeons have a special sense of direction.

In the last generation man has made machines which

perform the function of eyes, using rays which we cannot see. For our eye is a very limited organ. The vibrations of violet light, which are the most rapid that we can see, are only twice as many per second as those of red, the slowest.

Other animals can see rays which we cannot. Thus a bee can be trained to go to a glass of sugar water lit up by ultra-violet "light" which is vibrating too fast for the human eye to see. And within the ultra-violet it can distinguish different "colours." But it cannot see red light. So green leaves, which are green because they absorb red light, look like yellow to the bee, and hence flowers which attract bees are rarely yellow or green. On the other hand, many flowers which are white to us appear coloured to bees.

Photographic plates can be sensitised so that they "see" many rays which we cannot. The slowly vibrating infra-red rays can penetrate fog or haze which scatters ordinary light. So photographs can be taken through fog, and through the haze which occurs in small amounts on the clearest days. A good photograph of the French coast has been taken from an aeroplane five miles above London.

The ships and lighthouses of the future will probably shine infra-red lamps in fog, and every ship will be equipped with an "eye" consisting of a battery of detectors of infra-red rays looking out in different directions.

Ultra-violet light has a special value for the following reason. Even with a microscope, one cannot distinguish the shape of things smaller than a wavelength of light—that is to say, about $\frac{1}{60,000}$ inch. That is to say, things can only be magnified about 3,000 times with any advantage in clearness.

But if we use ultra-violet light we can detect the shape of considerably smaller bodies, and it is now being used in London by Barnard to photograph the germs of foot-and-mouth disease. Unfortunately, however, ultra-violet microscopes cost nearly as much to build as range-finders for heavy guns. And naturally at the present time we prefer to spend the money on range-finders.

The wavelength of X-rays is about a thousandth of that of ordinary light. They can be used for two purposes: to see visible things which are hidden, such as the bones and heart of a living man, and also to see things like atoms, which are too small to detect with ordinary light.

The method by which atoms are seen was invented by Sir William Bragg, and is as follows. If you clap your hands near a fence with regular slats you get a musical echo back. The reflected waves form a regular sequence with a wavelength twice the distance between the slats. Similarly, if fine parallel lines are ruled on a copper plate they reflect a colour depending on the distance between them, just as the fence reflects a certain note. Many of the beautiful colours of birds and beetles are produced in this way.

Now in a crystal the atoms are arranged in regular layers, and reflect X-rays of the right wavelength. If we could see X-rays, we should say that the crystals had iridescent colours like a pearl or a pigeon's breast. By using X-rays of different "colours"—that is to say, wavelengths—and twisting the crystals about, we can find how the atoms are arranged in them.

The results have been amazing. For a long time the organic chemists had made diagrams or models of chemical molecules, showing atoms arranged in chains or rings. Some chemists believed that these were essentially true pictures. Others said that they were only a kind of shorthand for expressing the chemical properties of things, and that we should never know what molecules were really like.

Others, again, went further still, and said that molecules and atoms were not real at all, but only mental constructions. Lenin attacked these latter in *Materialism and Empirio-criticism*. But he may have been too busy to learn how completely his criticism was confirmed during the last four years of his life.

The X-ray photographs have shown that the atoms are really arranged as the chemists had predicted, and that crystals have properties which we should call colours if our eyes could see X-rays. And the new knowledge of crystals has explained many strange things.

Goldschmidt of Oslo explained why some rare elements such as radium and platinum are found as impurities in common minerals because the rare atoms just fit the instertices of the crystals. Bernal of Cambridge used X-rays to determine the size and shape of disease germs too small to photograph even with ultra-violet light.

If the earth or the moon were a living animal it would perhaps have "eyes" sensitive to radio waves. They would consist of receiving sets each picking up waves from a certain direction, and tuned in to a certain wavelength or "colour." With these it would "see" broadcasting stations as lights, and also a source of short radio waves in the sky, emitting waves round 15 metres in length, and rising and setting like the sun.

This mysterious heavenly body, the sun of the radio

world, rises and sets 366 times, not 365, in the year. It is near the middle of the milky way in the constellation Sagittarius, and was only discovered two years ago, so we don't yet know much about it.

Just as lamps are coloured because they give out light of certain wavelengths, so non-luminous things are coloured because they absorb some wavelengths and reflect others. Thus gold is yellow because it absorbs blue light and reflects the rest.

In our upper atmosphere there are layers of gas which reflect long radio waves and let through shorter ones, so that a radio "eye" would see the sky as red in colour. These layers are being investigated by Professor Appleton and his colleagues with a transmitter at King's College, London, and a receiver at Hampstead which picks up the echoes from them.

If there were no reflecting layer, radio waves would fly off into space in straight lines, and it would be impossible to broadcast from Daventry to Glasgow, let alone Australia. When this was first realised it was thought that there was only one reflecting layer. Now we know that there are three, and that below the lowest there are little clouds of "coloured" gas.

Such are some of the real but invisible things around us. Materialists are often accused of reducing the human mind to the level of mere matter. But when one knows a little science, materialism comes to mean the belief that the mind is as wonderful as the mater world which it reflects.

NATURE'S SPEED LIMITS

IN September, 1938, Eyston once again beat the world's record for speed on land. So at present the record stands at 357·5 miles per hour. Only ten years ago the land record stood at the very modest figure of 171 miles per hour. Is it going to be doubled again in the next ten years?

In the nineteenth century many people believed in continuous progress. They thought that social conditions would go on improving steadily, just as they were then improving in England. And the physicists of that day never considered the possibility that there could be any upper or lower limit to the speed of a moving object.

An aeroplane makes a noise, partly due to the exhaust and the propeller, partly to its motion through the air. This sound travels away at 750 miles per hour in each direction.

So if a plane could move at 750 miles per hour it would just keep up with its own sound. But a sound is a disturbance of the air. So the plane would find itself in the centre of a violent and irregular hurricane, and would be knocked out of its course and perhaps torn to pieces.

We know that this is so from practice as well as theory. The bullets fired at the battles of Waterloo and Peterloo left the old muzzle-loading guns at about 500 miles per hour; those from modern rifles start at 2,000 miles per hour or over. But at speeds near that of sound, shooting is very erratic indeed. And near the speed of sound the resistance of the air increases very greatly.

So I doubt if planes will fly at 750 miles per hour or

over for a good many years to come. When they do so they will have to fly at a very great height, where air is thin. And just because of the thinness of the air, propellers will be useless, and designers will have to employ the same means of propulsion as a rocket, or perhaps some entirely new method.

This whole question is an example of a principle which holds in politics and economics as well as in physics, the principle called by Hegel and Marx the "transformation of quantity into quality." As the speed increases, the sound becomes first a trivial effect, then a slight hindrance, and finally a danger. There is a critical stage, but when this is past, things often become easier again. We can all think of examples of this principle in the field of politics.

Light travels about a million times quicker than sound. The fastest moving large things that we know are some very distant groups of stars, which seem to be moving away from us at about one-tenth of the speed of light. But in the laboratory we can easily make electrons or electrically charged atoms or molecules move at speeds very close to that of light if we apply a very strong electric field.

But although their speeds reach 99 per cent. of that of light, or even higher, no one has managed to get them to go as fast as light. There are good arguments to prove that no one will ever be able to make a particle move as fast as light, but they do not amount to absolute certainty. It is, however, quite sure that for practical purposes light gives us a speed limit which we are not likely to exceed for a long time.

Human physiology sets no definite speed limit, but if you are going fast, turning is dangerous, because it drains the blood into the parts of the body on the outside of the turn, and may cause fainting. In particular, if a pilot turns a rapidly moving aeroplane upwards, his blood is forced down into his legs, and he may faint.

Are there any lower limits of speed? At first sight the answer is "No," because we are all familiar with things at rest relative to the earth. But even so we know that the atoms in resting bodies are moving. It is this irregular motion of atoms which we feel as heat. When anything is cooled, it usually contracts, because this motion is slowed down. Now, Rutherford found that fast-moving particles mostly went through solid bodies, only a few being stopped by the cores of atoms. So if a body could be cooled down so that all the atomic motion was lost, it would contract so enormously that a weight of several tons could be carried in a matchbox. This does not happen, because each sort of atom has a minimum amount of motion which cannot be taken away from it. About twenty years ago physicists had to admit that matter and motion were inseparable. This is one of the basic principles of the quantum theory, and has worried some physicists so badly that they do not believe that matter or motion have any real existence.

However, it is quite a familiar idea to readers of Engels' *Anti-Dühring*, and although, of course, Engels did not predict the whole of modern physics, he did make it easy to understand how inextricably matter and motion are bound up.

A watch is a different sort of thing according to whether

it is going or has stopped, and a bullet is different according as it is going quicker or slower than sound. So a good Marxist will believe that men will some day travel 1,000 miles an hour, but not without a revolution in their means of propulsion.

ATOM-SMASHING

THE Press has recently been full of many accounts—mostly nonsensical—of atom-splitting in a number of laboratories. To understand them we must go back and study the history of this subject. The first person, so far as we know, to say that matter was made up of many different kinds of atoms—that is to say, indivisible particles—was Democritus, a Greek philosopher on whom Marx wrote a thesis for his doctor's degree.

But Democritus' theory remained a theory till Dalton of Manchester about 1808 showed that the composition of various substances can be explained if they are built up of molecules, each containing a small number of atoms. About thirty-five years ago the size of atoms was first accurately determined. It takes about 100 million placed on end to measure 1 inch. This, by the way, is not too large a number to imagine. It takes about 100 million small water drops to fill 1 cubic yard.

However, most of the space in a solid is empty, and a small, quickly moving particle will go right through it. The characteristic part of an atom, which cannot be divided by ordinary means, is called the "nucleus." It has a positive electric charge, and is surrounded by anything from one up to ninety-two negatively charged electrons, which move round it, and keep the space round it clear. But almost all the weight of a substance is in its nuclei.

In a chemical change, such as burning a match or digesting a sausage, the pattern of the electrons round the nuclei is altered, but that is all. And if these were the only sort of changes possible, the nineteenth-century chemists who either called atoms "the eternal bricks of the universe," or else thought that they had been created once for all by God, would have been right. And Engels, who no more believed in indestructible atoms than indestructible capitalism, would have been wrong. He died before Rutherford showed that atoms were not eternal.

Now when a nucleus alters, two things happen. It generally changes its electric charge, and therefore the number of electrons around it. Thus radium has a very loosely held electron, which can easily move from the neighbourhood of one nucleus to another, and it is therefore a metal, in which electrons can move freely and which therefore conducts electricity. But when its nucleus loses some of its positive charge it becomes a gas, radon, in which there are fewer electrons round the nucleus, but they are held much tighter; so radon will not conduct an electric current, even when made solid by cold.

Besides this the break-up of a nucleus always liberates a great deal of energy, partly in the form of rays, and partly by setting the bits into which the nucleus divides into very rapid motion. This was how the break-up of a nucleus was first detected.

The first nuclei which were discovered changing were those of the metal uranium, which has the heaviest nuclei of any atom. But soon radium and other radio-active elements were found whose nuclei are much more unstable than those of uranium. Then Irene Curie-Joliot, the daughter of the Curies who had isolated radium, found that by shooting particles from radio-active elements into the nuclei of stable elements they could be made to break up.

Rutherford's pupils, especially Chadwick, Cockcroft and Walton, showed that rapidly moving particles which could break up a nucleus could also be got by passing a million-volt current through hydrogen, and that the most effective of them all was a particle called the "neutron," with no electric charge.

Fermi in Italy bombarded uranium with neutrons, and showed that a few uranium nuclei could be changed into something else which broke up much quicker than uranium. He thought that these were heavier than uranium nuclei, in which he turns out to have been wrong. In all these cases, whether the radio-activity was natural or artificial, it appeared that a large nucleus, when it became unstable, spat out a small particle, or perhaps several in succession.

However, in 1938 Hahn and Strassmann in Berlin found that they could not separate one of the products of uranium from barium—that is to say, it had the same chemical properties, though its nuclei were radio-active. Hahn's former colleague, Lise Meitner, who was Germany's greatest woman scientist, but has had to leave, is working in Sweden, and along with Frisch in Denmark, she saw that this meant that the uranium nuclei had been split into two nearly equal parts.

This theory was immediately confirmed in Copenhagen, and has since been shown to be true in about a dozen laboratories in America, England, Holland, France and Germany. Apparently a uranium nucleus, when hit by a neutron, breaks up into a nucleus of xenon (a rare gas) and another of the metal strontium. These latter break up in the usual way by throwing out small fragments only.

The energy disengaged in this reaction has not yet been accurately measured, but it is far greater than that involved in ordinary radio-activity. In fact, if one could make neutrons go where one wanted, it could be used as a source of power, and uranium would give out some millions of times the energy of the same weight of coal, though probably rather slowly. However, at present neutrons are so hard to make and so impossible to guide into a desired path that this is not so.

Nor is it likely to be for a generation or more.[1] However, there are physicists who believe that within this century artificial radio-activity will be used as a source of power. If so it will revolutionise human society as completely as did the steam engine, for it will be possible to set up anywhere an engine which will work for centuries without fresh fuel. If that day ever comes, Hahn and Strassmann's discovery will prove to have been an important step towards it.

[1] Later work has shown that it is at least possible within the next few years.

THE DISORDER OF NATURE

PRIMITIVE men, so far as we can understand their thought, have no idea that Nature is orderly. They generally believe that natural events are due to all sorts of spirits. This storm was due to the anger of such and such a god. That man died from a wizard's spell. The good crops last year were caused by a successful bit of magic. And so on.

Gradually men learned of the orderly processes in Nature. The priests in Egypt learned that when a particular star was first visible before sunrise the fertilising flood of the Nile was due. And by this and similar knowledge they controlled the whole people. The properties of different minerals were discovered, first by craftsmen, and later by scientists. Natural objects were neatly classified in species, families and orders.

They were even arranged in "kingdoms," the animal, plant and mineral kingdoms, in analogy with human states. Man came at the top of the animal kingdom, then other vertebrates formed a sort of nobility, the mammals being dukes, while fish were barons, so to say. There was a middle class of insects, and a proletariat of worms, molluscs and other lowly creatures.

Towards the end of the eighteenth century the world was regarded as a great machine. Some people thought that it had been designed by God with a special eye to the needs of man. Others thought that it was the product of eternal laws. But everything was neat and tidy, if we knew enough.

Things looked even better in the nineteenth century

when all the known kinds of matter were found to be made up of just ninety chemical elements neatly arranged in compounds. Darwin's teaching that animal and plant species were not fixed was awkward for a moment. But then that was fitted into the scheme, too. Instead of the animals being arranged in kingdoms according to a Mediæval conception, they were competing just like nineteenth-century capitalists before the days of the great trusts.

This sort of science is still taught. But it is out of date. Mme. Curie and Rutherford began the revolution when they found that some kinds of atoms were unstable. And now every month new unstable kinds are described, some only lasting for a fraction of a second on the average.

Naturally enough, these unstable ones are rare. But it turns out that those which we thought were the eternal bricks of the universe are merely the more lasting sorts. Probably none of them last for ever. At bottom Nature consists of processes, not things.

Worse still, there is no way of predicting just how even a stable atom will behave in the next second. The reason is not because atoms have free will. That theory is merely a reversion to the ideas of primitive savages. There is no ground for thinking that atoms have wills at all. The reason is this. Nearly 100 years ago Marx said that men never merely observe Nature. They always change it. Now when you observe an atom you can't help moving it, for even light exercises a slight push on things. So your observation always changes the thing observed.

And you can't tell how much it has been changed till you observe it again, and by doing so change it again. It took the scientists a long time to prove that Marx' principle held in physics as it does in society. But when they did they found it very useful.

For although you can't predict what one atom can do, you can use the uncertainty principle to predict what a lump of matter consisting of millions of atoms will do with very great certainty. Just in the same way we can predict that unemployment will increase during the next few years,[1] though we can't say whether John Smith or Bill Jones will lose his job first.

Just as the atoms are disorderly, so are the stars. We live in a fairly neat part of the universe where the planets go round our sun very regularly. But many parts are crowded with stars, like the Pleiades, where any planets would soon be thrown out of their orbits. Others are full of dust which would slow them down.

We happen to live in an orderly system, and that is why life has been able to evolve on the earth. But to say that it has been made orderly for our benefit is putting the cart before the horse, like saying that it is very lucky that so many towns are on navigable rivers and so few in deserts or on mountain tops. The rivers are responsible for the towns, and not the other way round.

Even in our solar system the earth is probably the only planet suited for life. Jupiter is too cold, Mars too dry, Mercury has no air or water. Venus may possibly be inhabited.

Living creatures are not so perfect as was once thought. And it is only their imperfections that make evolution possible. If they did not die they could not make room

[1] The slump should be at its worst about 1942 or 1943, unless there is a war.

for anything better. If there were no exceptions to the laws of heredity they would resemble their parents exactly, and evolution would be impossible, just as there would be no progress in a society where everyone fitted in perfectly and always obeyed the laws.

It is no good saying that we should go back to Nature. Wild animals have plenty of pain and disease. The rabbits in Canada are wiped out by an epidemic every eleven years. Then the foxes which eat them starve, and large numbers of foxes are caught in traps. Primitive men live shorter lives than civilised men. We have certainly gone too far from Nature in some respects. But we can't go back all the way.

The fact is that Nature is a strange mixture of chance and necessity. Many writers to-day give vent to the gloom caused in them by the decline of capitalism by saying that man is the slave either of necessity or chance. The opposite is true. As soon as we really understand a law of Nature, we find a way of using it or even avoiding it.

Let us take a simple example. It is a law of Nature that if we drink water contaminated with typhoid bacteria we are likely to get typhoid, as in the Croydon and Somerset epidemics. But the moment we understand this law it becomes untrue, because we can add "but not if we boil or chlorinate the water before we drink it."

In the same way we suffer every ten or eleven years from an economic depression causing widespread unemployment. Yes, but not if we understand the economic laws which cause these cycles, and act on our knowledge. There was no slump in the Soviet Union in 1931, and there is no sign of a coming slump there to-day.

If Nature or society is not as orderly as we should like, it is up to us to make it so. But we can only do so if we recognise the disorder, investigate its causes, and act on our knowledge.

MACHINES IN SCIENCE AND INDUSTRY

SOME machines are first used in industry and then in science. For example, the windmill was used for grinding corn, and then a small-scale windmill called an "anemometer" was used to measure the speed of the wind. Sometimes things are the other way round. Thus most of the instruments ending in "meter" were originally used by scientists for measuring.

The thermometer and barometer are still used for measuring by ordinary people. But the gasometer, which was originally a laboratory dodge for measuring gas over water, is now used for storing it. More usually there is give and take. The X-ray tube was first a laboratory invention, then used in medicine and industry. It was so much improved and cheapened that it came back to the laboratory hardly recognisable.

An apparatus which has escaped out of the laboratory into industry in the last generation is the centrifuge, or cream separator, and one which will escape soon is the ultra-centrifuge. In the ordinary centrifuge we can put a liquid containing particles lighter or heavier than itself —for example, blood or milk. The liquid is put in a cup which is spun round, and the heaviest part flies outward. For example, in the case of blood, the red corpuscles are thrown outwards because they are heavier than the rest. But with milk the drops of cream, which are lighter, are left behind in the middle, while the skim-milk goes to the outside.

Still more powerful separators will bring down bacteria. So if we had a sufficiently strong one, the Swedish physicist Svedberg argued, we could separate sugar or any other dissolved substance from water, for the dissolved sugar consists of particles, which are a lot smaller than bacteria, but still have a perfectly definite size and weight.

Svedberg began on fairly large molecules, such as those of hæmoglobin, the red substance in blood. His latest type of ultra-centrifuge is a wheel about 2 inches in diameter. It is supported on horizontal bearings, and spun by twin oil turbines at each end of the shaft. It spins in an atmosphere of hydrogen at a pressure of about $\frac{1}{40}$ atmosphere. At this low pressure the gas offers little resistance, but carries off the heat made by the turbines and bearings.

This wheel can spin round anything up to 140,000 times per minute; so its edge is moving rather faster than the fastest aeroplane. And if a liquid is put in special cells bored in this wheel the separating force is 700,000 times that of gravity. Naturally enough, the whole thing must be enclosed in armour plate, for if it flies to pieces it may cause a bad accident.

With this ultra-centrifuge the size of protein molecules has been accurately determined, and the half-living chemical agents of diseases such as vaccinia (cowpox) in man and mosaic disease in tobacco have been prepared in a pure state, although their particles are far too small to be seen with the microscope.

Another kind of ultra-centrifuge was invented by Henriot and Huguenard in France. This is shaped like a "put and take" top, and nearly, but not quite, fits into a conical hole. It spins round on a film of air or steam blown out obliquely in jets through the hole. These tops are very much cheaper than Svedberg's apparatus, and have been used for a number of purposes. For example, a mirror placed on one of them will give over a million flashes per minute, and can be used for ultra-slow motion-pictures of electric sparks.

But the most amazing development is the axial magnetic suspension invented by Holmes and Beams of the University of Virginia. Its axis is a steel needle which is hung below a solenoid—that is to say, an electromagnet without a core. Its position is regulated by a beam of light which is cut off if the centrifuge falls below a certain level, and thus turns on more current. The whole thing is in a high vacuum to abolish air friction, and is set spinning by a magnet on a Henriot top.

Here at last is something like a frictionless bearing. Not quite frictionless. But when left to spin it only lost one five-thousandth of its speed in a minute. So it would take about two days to lose half its speed. Compare this with any other kind of bearing, and you will have little doubt that the principle will some day be applied in industry.

These centrifuges are already being applied in chemistry. Chlorine is a mixture of two elements whose atoms weigh thirty-five and thirty-seven times as much as a hydrogen atom. Their chemical properties are so close that there was no way of separating them. But this can be done, though not yet completely, by spinning.

It is quite possible that in another fifty years spinning will supersede chemical methods and those depending on heat for separating substances in industry. For example, in distilling spirits, we make use of the fact that alcohol boils more easily than water, and therefore comes off first when a mixture is boiled. It may be more economical to use the fact that alcohol is lighter than water, and can be spun out of it as cream is spun out of milk. In the same way salt can be largely removed from sea-water.

We are still a long way from such industrial applications, and monopolists will probably hold them up for some time when they become economically possible. However, within the lifetimes of some of us the industrial ultra-centrifuge and the frictionless magnetic bearing should cease to be laboratory curiosities and become part of the technique of large-scale production.

EVOLUTION AND ITS PRODUCTS

IS DARWINISM DEAD?

WE are often told that Darwinism is one of the exploded myths of the nineteenth century. Some people say that evolution is a lie, others that it was determined by a life-force or a creative urge, or something vague of this kind, and not in the common-sense manner that Darwin thought.

Now, anyone can find mistakes in Darwin's works. He was not infallible. In the same way Newton's account of gravitation was not quite right, but it is still good enough for the purposes for which it was framed. Dalton's account of atoms has had to be revised, but it is still the basis of chemistry. And although a few of Marx' predictions have not come off, he gave an incomparably better account of future changes in society than any of his contemporaries.

When Darwin wrote we knew much less about animals and plants of the past than we do now. He said that the complicated forms living to-day were descended from simpler forms in the past, and that intermediates would be found. If they had not been found, he would have been proved wrong.

But they have been found. The bones of at least six

different kinds of animal intermediate between men and apes have been discovered. Some of them used fire, so they are best thought of as men; though if they were alive to-day they would probably be shot as big game by sportsmen. Others probably had no industry, and were apes rather more man-like than the gorilla.

Many other links have been found. For example, there were lizards with teeth like those of mammals, and birds with teeth and long, bony tails. Intermediate between fish and newts were creatures with stumpy legs and nostrils, like those of many fish, on the undersides of their heads. And between ferns and flowering plants there were ferns carrying seeds. Only in 1937 creatures intermediate between lampreys and other fish were first described.

And these creatures occur in the rocks at the depth where they were expected. Darwinism would be exploded to-morrow if the skeleton of a man or a horse were found embedded in a coal seam, where only the bones of animals like newts and lizards are known. In the same way, Marxism would be disproved if a fascist state raised real wages all round, abolished unemployment and reduced its armed forces. But neither of these things has happened.

The real test of the evolution theory is practice. The date of a rock can be determined from the fossils in it. And on the basis of this dating predictions are made about the minerals to be found in it. Indeed, the details concerning Foraminifera, microscopic sea animals whose skeletons are found in limestone, are among the jealously guarded secrets of the great oil trusts. They are of such value in predicting where oil will be found that they have to be kept from rivals.

Men and women who do not believe in medical science have the courage of their convictions, and go in for Nature cures, osteopathy, Christian Science and so on. Rich people who say that they disbelieve in evolution do not invest their money in mining ventures undertaken contrary to the teachings of geology, which is based on the theory of evolution. When a company of anti-Darwinians start looking for gold in Kent or coal in Cornwall I shall take their doubts more seriously.

Besides giving an account of evolution, Darwin proposed a theory of why it had occurred. The main trend had been determined by natural selection. Animals and plants varied, the variations were partly inherited, and since some variations allowed their possessors to leave more descendants, they spread through the population, and thus the population changed, and a new species was formed. For example, some animals have thicker hair than others of the same species, and the differences are inherited, as anyone can see by comparing fox-terriers and collies. The hairier animals will do better in the Arctic, and the less hairy in the tropics. And thus two races will be formed which will later develop into two species.

This theory was criticised from many angles. Some people think that habits acquired during life are handed down to the descendants. This was believed by Lamarck, a French scientist who had anticipated many of Darwin's views on the historical side of evolution. But no one has been able to prove it experimentally, and many facts of natural history speak against it.

No animals have more complicated instincts than

worker bees. They do not reproduce, and are not descended from other workers, but from queens and drones. So if habits are inherited, they should long ago have lost their instincts, which enable them to build honeycombs and to do other remarkable feats. They should have come to behave like queens or drones.

Lamarckism is now being used to support reaction. A British biologist who holds this view thinks that it is no good offering self-government to peoples whose ancestors have long been oppressed, or education to the descendants of many generations of illiterates. He has, however, to explain why even the children of orators must still be taught to speak, though men have been speaking for hundreds of generations.

Darwin's theory of natural selection by the survival of the fittest has also been used to defend human injustice. The facts, however, are against this attempt. Throughout history ruling classes have exterminated themselves. At present the rich in England leave fewer children than the poor, even when allowance is made for the lower infantile death-rate of the rich. So, from a Darwinian point of view, the poor are fitter than the rich. The capitalist may win in the struggle for cash, but the workers are winning the struggle for life.

In the same way the most successful species have not usually been the meat-eaters. Rabbits are commoner than foxes, and the huge armoured reptiles of the past were replaced by smaller and more intelligent mammals and birds. Above all, co-operation, or rather an emotional make-up which allowed co-operation, has been a great factor in fitness.

Anti-Darwinians still say that species are quite different from varieties, because varieties—for example, greyhounds and bulldogs—although they look different, can be crossed together, whereas species—such as the horse and donkey—usually cannot be crossed, or give sterile hybrids like the mule. However, artificial species, which cannot be crossed, have lately been made from a single original species. This was first done about twenty years ago with tomatoes in London, and more recently with flies in Moscow.

Darwin's views on variation and heredity have had to be greatly modified, and his account of natural selection was a good deal too simple. Nevertheless, modern biology is built on foundations which Darwin laid.

SOME MISSING LINKS FOUND

THE phrase "missing link" was coined about fifty years ago for a fossil form which enthusiastic Darwinians hoped would convince everyone that men were descended from apes. Since then different workers have found at least six different types of skull, and sometimes complete skeletons, which are more human than those of any living ape, and more ape-like than those of any living man.

Whether they should be called bones of apes or men depends not so much on their anatomy as their habits. If they had started real production, as opposed to mere collection of food, they were definitely on the path which leads, through increasing improvements in technique, up to civilisation. Making a fire must certainly be regarded as a form of production, and one of these extinct forms, *Sinanthropus pekinensis*, of which a number of skeletons have been found near Pekin, used fire.

Some of the others, though no more ape-like in build, may have been apes in behaviour. I confess that I am not much excited by the anatomical details of these early men. We learn much more about them by studying their tools than their bones. Certainly men chipped stones for an immense time before they began to paint on cave walls or scratch on bones, unless we have been singularly unlucky in missing all traces of early art.

And I am more interested in some of the other links in the evolutionary story, bridging far greater gaps than that between apes and men. All ordinary fish have a pair of jaws. A few have no jaws, but a round mouth like that of some worms. The best known in England is the lamprey, which lives in river mud. These primitive fish also differ from all modern ones in having no paired fins, such as have developed into our legs and arms; and those that live in the sea have blood which is very like sea-water, whereas in modern fish the blood has been greatly modified.

Skeletons of these fish are found in rocks such as the lower Silurian of Wales where there are no fish with jaws. Quite recently Professor Watson of London has described another group of fossil fish which bridge the gap between them and the modern fish, having a very simple lower jaw and the beginnings of paired fins.

Another gap which is being filled by the study of fossils is that between fish and amphibians such as the newt. If anyone doubts the possibility of fish coming out of the water and developing legs out of fins, he had better look at the mud-skippers in the farthest section of the aquarium of the London Zoo. They spend most of their time on land, and have developed a sort of elbow in their front fins with which they hop.

Unfortunately, they are about 250 million years late in their attempt to colonise the land from the water. Our own ancesters did it at the time when the Old Red Sandstone was laid down, and we can trace the stages of the formation of limbs from fins, and see how the nostrils, which were originally underneath the head, moved up to the top, where they are required in animals such as frogs or crocodiles which breathe air, but spend much time in shallow water.

All these links were unknown in Darwin's time, though, if evolution was a true theory, they must have existed. Another wonderful set of connecting forms has been found in South Africa, whose rocks contain not only gold and diamonds, but the bones of hundreds of different animal species which link reptiles with mammals. For example, some are like lizards or crocodiles, but have their bodies lifted off the ground, and teeth specialised into cutters, dog-teeth, and grinders, whereas reptiles generally have all their teeth alike.

In the same way, intermediate stages in the evolution of society have been discovered among primitive peoples. In *The Origin of the Family*, Marx' colleague, Engels, described a stage where property was held in common by a clan, and a woman always married outside her own clan. So a man's bow and arrow, or garden, did not pass to his own children, who belonged to his wife's clan, but to his sisters' children who belonged to his own.

If property was to be passed down from father to son, the primitive communism of this system had to be broken up. Anthropologists have recently described a stage in the break-up of primitive communism which Engels had not suspected. In order to keep property in the family, cousin marriage is made compulsory, or at least usual, so that a man's property can be handed down to his children, not because they are his, but because they have married his sisters' children.

Naturally critics say that Engels was wrong because he did not describe all the stages between primitive communism and private property. This is as if one blamed Darwin for not describing fossils connecting the great groups, which were unknown in his time. Research has certainly modified the conclusions of Darwin and Engels in detail, but it has confirmed the general accounts which they gave of the evolution of animals and of societies.

LIVING FOSSILS

A FEW weeks ago[1] a most peculiar fish was caught off Port Elizabeth in South Africa. It looked no odder than many other deep sea fish, though any observant person would have noticed that it had a small extra tail sticking out of the middle of its ordinary tail, and that its paired

[1] In February, 1939.

fins, instead of consisting of a fan of spines, had something resembling a stumpy limb in their centre. You can see rather similar fins in the lung-fish at the Zoo in London.

Its peculiarity was of another sort. It belonged to an order of fish called the Crossopterygii, which were common enough in the swamps where the coal seams were formed, but of which no fossils had been found in strata later than the chalk. In fact, they were supposed to be as extinct as the great reptiles which once lived in most parts of the world.

This particular group of fish is much closer than any other group to those which are believed to have come out of the water during the Devonian or Old Red Sandstone period, and to have been the ancestors of four-footed land animals, birds and men. Their bones were known already, but a study of their soft parts, and particularly their heart, brain and swim-bladder, will be of great interest to students of evolution.[1]

And this particular discovery will be welcome for another reason. The ancestry of some animals—for example, the horse—is very well known from a study of fossil skeletons. But there are some serious gaps in other lines. For example, there is little doubt that birds were descended from reptiles. And a few primitive birds have been found, with long bony tails, claws on their wings and numbers of teeth. But they already had feathers, and nothing is known of how they originated from reptiles.

Darwin and his followers always stressed the imperfec-

[1] Unfortunately the workers in the local museum who prepared the skin and skeleton of this fish threw away the soft parts without proper study, probably because they were "high." However, other workers are now busily fishing for more specimens.

tion of the geological record. That is to say, they said that only a very few of the millions of animal species that have lived in the past have left a record which has so far been discovered. Their opponents tried to make out that this was not so, and that there were therefore gaps in evolution which could only be explained by new creations —for example, of birds.

Here is a case of an animal whose ancestors must have lived in the sea for about 50 million years since the Chalk Age, but none of their bones have yet been found. So in its case there was a 50-million-year gap in the record. Another reason for rejoicing among palæontologists is this. The bones and scales of fossil fish are generally found somewhat crushed, and a reconstruction demands some imagination. Besides which, on the basis of evolutionary theories, palæontologists had said what their hearts and other soft organs must have been like.

Now here is a wonderful chance of checking these theories by actual dissection. A scientific theory is a mere string of words unless one can check it in some such way as this. So is a political theory. For example, Marxist theory predicted that Mr. Chamberlain would back Hitler at Munich, while non-Marxist theory predicted that he would back Britain. The *Daily Worker*, being the only Marxist daily paper, was the only one which forecast the result correctly.

Many people ask how it is that, if the evolution theory is true, a fish remains almost unchanged for so many millions of years. One reason is this. Many of the very primitive animals which are alive to-day mature very slowly. For example, the Australian lung-fish is very close indeed to some ancient fossils. But it takes at least twenty years to mature, and perhaps an average generation is fifty years. So a fish which breeds when a year old has had fifty times as many generations to evolve.

The newly discovered fish must be a slow grower, for it is about 6 feet long, and lives in the middle depths of the sea, where food is scarcer than at the surface or on the bottom. In just the same way trees are generally more primitive than small plants, because they have not had so many generations to evolve. The most highly evolved plant orders, such as composites like the daisy, grasses, and labiates like the dead-nettle, are almost all quite small; while most trees have cones like the pine, or flowers of a simple type like those in catkins.

There is a law of uneven development in animal and plant evolution as in the social evolution of capitalism, and plenty of animals and plants have not changed very much for an enormous time. There is no general law which makes either animals or societies improve in all cases. According to Darwin, animals improve through a struggle, and according to Marx societies do so through a different kind of struggle.

While it is unlikely that many large land animals which were thought to be extinct will be found, there is one interesting possibility. The *Chalicotheria* were a very odd group related to horses, tapirs and rhinoceroses. They had a head rather like a horse, but claws instead of hooves. Their fossils are known up to the time of the Ice Age, and they are generally thought to be extinct. But in the forests of northern Kenya the natives report an animal about the size of a bear which answers fairly well to this description. No European has ever seen one close, but pieces of skin belonging to no known living species have been seen.

However, perhaps the most interesting "living fossils" are invertebrates, quite small animals which survive in out-of-the-way corners. Some of them are geographically isolated, like *Anaspides*, a primitive kind of lobster with joints all the way down its back, instead of having a carapace like a Life Guardsman. This is only found in a few lakes in Tasmania.

But in case you think you must go abroad for such creatures, it is worth remembering that in 1866 Sir John Lubbock (the originator of bank holidays) discovered *Pauropus* in his own kitchen garden. It is a lively little white beast 1 millimetre long with seven pairs of legs, and something like the ancestors of insects. So any keen naturalist may make as big a discovery to-morrow as the South African fish.

BEYOND DARWIN

DARWIN taught that the direction of evolution was determined mainly by the survival of the fittest. Animals of the same species differ among themselves. One mouse can run faster than the average. Another has better hearing. Still another has sharper teeth. These differences are at least partly inherited. And as there is not room in the world for all the mice born, the fittest on the whole survive, and thus the species gradually changes.

A character which is useful in one environment may be harmful in another. Thus thick fur is useful in the Arctic and harmful in the tropics. Wings are generally useful to an insect in the middle of a continent, but dangerous on small islands in the ocean, where winged insects are blown out to sea, but wingless ones survive. This is one of the ways in which a species divides into two or more new species.

Marx and Engels accepted this theory of the struggle for life "as the first, temporary, incomplete expression of a recently discovered fact." They pointed out that "Darwin discovers among plants and animals his English society" based on unrestricted competition. And, of course, since Darwin's time many theorists have tried to justify cut-throat competition and the oppression of the weak in the name of Darwinism.

But Marx and Engels did not deny the struggle for life in Nature because they thought that men could and should behave better than animals. Kropotkin wrote of co-operation in Nature even between different species. This occurs, but it is exceptional.

The dialectical method in science is to push a theory to its logical conclusion, and show that it negates itself. For example, we know that the so-called atoms of chemical elements are not really indivisible. But this would never have been discovered if chemists had not believed in the existence of atoms, and investigated their properties with great care. Dalton's atomic theory is still the basis of chemistry. But it is such a good theory that it disproves itself, and makes way for a nearer approach to absolute truth.

It is the same with Darwinism. Animals and plants are not quite such ruthlessly efficient strugglers as they would be if Darwinism were the whole truth. It is true that a lot of what at first seems useless beauty is part of the struggle. Thus flowers are useful to plants because they attract insects. And they are beautiful to us because we share the æsthetic preferences of insects to some extent.

However, it has recently been shown that the struggle for life defeats itself if it is pushed too far. So long as a species is mainly struggling against other species or external nature, it usually becomes fitter. But when the struggle occurs within a species this is not so. Thus if male animals fight for females, the most successful fighters will have most children. So the species may develop weapons and instincts which are only useful in fighting their own kind.

In particular, mere size is an advantage in such struggles. Animals where the male is much larger than the female, such as the domestic fowl, the sea elephant, and many species of deer, are generally polygamous. Whereas in animals with monogamous families, such as most birds, the sexes are generally of the same size. And the study of fossils shows that a steady increase in size generally ends in extinction. Large animals are usually less fit than small ones for flying, burrowing, making their way through thick vegetation, walking on boggy ground, and in many other situations.

The Soviet biologist, Gause, has studied the struggle for existence among small animals and plants in aquarium tanks. If we put in two species one of which eats the other, the eaters increase until their prey diminishes in numbers, and then begin to die of starvation. The numbers fluctuate periodically, like the numbers of men employed under capitalism.

And a crisis may become so acute that first the eaten and then the eaters die out. To prevent this, it may be necessary that the prey should have some kind of shelter where the eaters cannot find them. In fact, too great efficiency may lead to extinction. It is very unusual for a herbivorous animal to eat up its food plants completely. But this sometimes happens, as when a plague of caterpillars completely strip the oak-trees in a region, and the caterpillars then die of starvation.

The Oxford biologist, Elton, has given many examples of this principle. For example, the Red Indians in Labrador used to hunt caribou with spears and other primitive weapons. When they got guns they killed off so many deer that they starved or were compelled to buy imported food and live in settlements, where they caught European diseases.

Elton takes the view that it does not always pay a species to be too well adapted. A variation making for too great efficiency may cause a species to destroy its food and starve itself to death. This very important principle may explain a good deal of the diversity in Nature, and the fact that most species have some characters which cannot be accounted for on orthodox Darwinian lines.

Elton is not, so far as I know, a Marxist. But I am sure that Marx would have approved of his dialectical thinking, and that it is on such lines that the Darwinian theory will develop. I do not think that Darwinism will be disproved. But it will certainly be transformed.

THE SMALLEST COMMUNISTS

MAN is not only a social animal, but a producing animal. Some animals—for example, many monkeys, wild cattle, and even fish which swim in shoals—are social, but do not produce. Others, like the spider with its web and the caddis-worm with its house, are constructive but not social. Some animals are constructive at one time and social at another, like swallows, which build their nests in pairs, but migrate in flights of many thousands.

There are, however, a few insects which practise social production, including the division of labour. These include ants, wasps, bees and termites. During the nineteenth century they were studied as individuals or small groups by naturalists such as Fabre and Lord Avebury. A good deal was discovered about their senses and powers. But no one knew how they communicated, and some people even held mystical ideas about a single soul shared by a whole bee-hive or ants' nest.

Now, you don't know much about an individual man or woman till you know about the sort of society to which they belong. Just the same is true of bees. Many people had kept glass-fronted beehives, but the first men to study one thoroughly was von Frisch, now Professor of Zoology in Munich, and his colleague, Rösch.

A beehive contains a queen, who is an egg-laying female, and a lot of workers, who are females but do not normally lay eggs. And during the summer young queens, who may found new hives, or replace the old queen, are produced, along with males called "drones," to mate with the young females. So a beehive is really a very large family, with its mother.

Von Frisch had the great and simple idea of marking every bee in a small hive with spots of paint on its back. The first marked bees were killed by their comrades because they had a foreign smell. So von Frisch smeared them with honey, so that by the time they had been licked dry they had the smell of their own hive.

As soon as even a few bees were marked, he made a dreadful discovery. The bee had been held up as a model of constant work. He found that while there was always some work going on, each individual spent a great deal of time sitting on the honeycomb and apparently gossiping with friends. Then he gradually worked out the life-history of a bee.

The queen lays an egg in one of the cells of the honeycomb. Those cells prepared as nurseries do not contain honey, but a special food made largely from pollen. According to the quality of the food, a worker or a fertile female is produced. A grub hatches out of the egg which first eats the food store, is later fed by the young workers, and finally hardens into a chrysalis.

The young worker emerges from this fully grown, and is immediately stood a drink of honey by a comrade. Within half and hour it is at work cleaning out its nursery cell, and other cells of the comb. Its next duty is nursing. At first it merely sits on brood cells and keeps them warm. Later it feeds the grubs, its younger sisters, with pollen and honey. Finally, it feeds them with fluid from a pair of special glands, as a female mammal feeds her young with milk.

At the end of the nursing period it makes its first trial flight, and then takes part in general indoor work. Honey and pollen are taken over from the bees which have gathered them, and stored in the comb. New comb is built with wax from special glands which act at this time. Needless to say, there is no private property, either in food or wax. Nor are there class or even craft distinctions. The difference between workers, queens and drones is fixed at "birth," like that of the human sexes. And the drones, who do not work, are massacred when they have done their one duty of mating.

During this time some more trial flights are made, and dead bees and rubbish are taken out of the nest. Then follow several days of sentry duty at the door. Some sentries are very fierce, and attack wasps and foreign bees. Others seem to be pacifists.

At about three weeks old the bees begin visiting flowers and bringing home honey and pollen. So far there has been no specialisation, each bee taking various jobs in turn. Later they specialise, each bee visiting one particular type of flower, and usually gathering either honey or pollen, but not both.

By experiments on feeding the honey and pollen gatherers, von Frisch not only tested their senses of colour, form, taste and smell, but discovered three words in their language. Two of these words are dances, and one is a smell. If a bee finds a lot of honey in a flower, she executes a special dance on the comb when she gets home. Other bees join in, and each in turn smells the dancer with its feelers.

Those which have specialised on the same flower—

say, snapdragon or cherry—recognise the scent and fly out in all directions. The more honey has been found the longer the dance lasts, and the more other bees go out. A different dance is used as a signal by pollen gatherers.

A bee which finds a large store can also open a gland on its back which lets out a sweet smell and acts as a general call to all bees from the same hive flying near. If a layer of varnish is put over its opening they cannot call their friends.

Probably future research will disclose more "words," and we may discover how the builders co-operate to make the comb, and how it is decided that the hive needs some young queens. Or perhaps our next steps will be with the language of ants or wasps.

It is sometimes said that bees are unprogressive because they are communists. This would be a fair criticism if individualistic insects were progressive. But dragonflies do not do differential calculus nor do beetles make sculpture. Social insects progress slowly because their brains are small and their societies limited. Bees from different hives never co-operate and sometimes fight.

But even in their communist society the bees are not mere machines with no individual character. This is particularly shown by their behaviour as sentries. And although the worker is worn out after five or six weeks' work in summer, it has had a varied life, including many different kinds of work, and a good deal of leisure.

EELS

Last Saturday[1] a group of the unemployed hooked a number of eels in the trenches which had been dug for Air Raid Protection on Primrose Hill. It is, of course, a great mystery how these eels got there. But the whole story of eels is mysterious, and it is with other mysteries that I shall deal in this article.

The eel is a fish which has lost its hinder fins, as snakes have lost both pairs of legs and men their tails. But it has a much more surprising deficiency. In all other fish, when full-grown, you can easily find a roe—a hard roe full of eggs in the female, and a soft roe which yields sperm in the male. But the eel is apparently almost sex-less. There is a mere rudiment of a gonad or sex gland, and only an expert can tell males from young females.

Thirty years ago the life-history of the eel was only known in fresh water. Where a small stream goes down over rocks directly into the sea, one can sometimes see, from January to April, a swarm of baby eels or elvers swimming up out of the sea. They are generally a couple of inches long and as thick as a stout string. They are extraordinary climbers, and can get up vertical rocks if they are covered with moss.

In large rivers they sometimes form such dense swarms that they have been caught quite literally in tons—an incredible waste of future food. They settle down, preferably in deep and muddy water, and grow for five

[1] February 11th, 1939.

to twenty years. They are hardy and adventurous creatures, especially the females, and will creep out of the rivers at night through deep grass into ponds. In the days before the water supply of towns was properly filtered, they used to swim up pipes, and were sometimes found blocking taps on the upper floors of houses.

They burrow into the mud and sleep there through the winter; and feed on worms, insect larvæ and small fish during the rest of the year. Females may grow up to 4 feet in length and 25 lb. in weight, though males rarely exceed 18 inches. But, however long they are kept in fresh water they never become mature. When fully grown, they swim off downstream to the sea in autumn, and after this nothing was known of their life-history until 1896.

There were many theories. A Scottish naturalist even believed that they were engendered from water-beetles, probably because he mistook parasitic worms in these beetles for young eels. The problem was solved, at least in part, by the great Danish zoologist, Schmidt.

If you drag a very fine net through the North Atlantic you sometimes catch a small transparent flat fish about the size and shape of a willow leaf or smaller. In 1896 the Italian, Grassi, had discovered that these fish never grew beyond 3 inches long and then change into young eels, as tadpoles change into frogs, losing three-quarters of their weight in doing so.

As Schmidt sailed about the Atlantic he found that the transparent eel larvæ got smaller as he approached an area beginning between Bermuda and the West Indies, and extending east for about 600 miles. Here the sea is over four miles deep. The eels from an area ranging from Iceland to the Canary Islands and from the Azores to Cyprus swim across the ocean to this abyss, becoming sexually mature as they go. Here they mate and die. This behaviour is found not only in heroes and heroines of Wagner's operas, but in other migratory fish. European salmon have the opposite habit to the eel. They do most of their growth in the sea, and swim up the rivers to mate, after which some of them die, but a great many swim down to the sea again, and come back next year. But the salmon of the Canadian Pacific coast, which we buy in tins, invariably die after mating.

How do the little eels find their way across the Atlantic? The answer probably is that they don't. They are carried by the Gulf Stream and are three years old when they reach Europe. Schmidt argued as follows. Eels are found in Western Europe from the Adriatic to Scandinavia. If the eels derived from British parents came back to Britain and those of Italian origin to Italy, there would be different local races of eel.

Now the number of vertebræ in the backbones of many fish vary. And Schmidt showed that in trout and other fish the average number varied in different local races, and was partly hereditary. But in eels from all parts of Europe the average was exactly the same, somewhere between 114 and 115.

On the other hand, the fresh-water eels of the West Indies and those parts of North America which drain into the Atlantic have a breeding ground close to that of the European eels, and have a different number of vertebræ, usually 106. So there are two quite different races or species of fresh-water eel, besides, of course, salt-water eels, such as the conger.

Near the Bermuda Islands the young of both sorts of eel are found together. They are sorted out, not by homing instinct, but in a simpler way. The American race become elvers, with a desire for fresh water, when they are only one year old and the swarm of little fish is still near the American coast. The Europeans do not change until they are three years old, by which time they are mostly near Europe.

How the adult eels find their way back to their place of birth and death is quite unknown. Possibly by memory, and perhaps by means of some sense which men do not possess, though many birds possibly do. Besides this problem, there are plenty more to answer. Some day a ship equipped with a special trawl to work at a depth of five miles will go to the eels' breeding ground, and we shall find out what an adult eel looks like. Later on we shall discover whether cold salt water is enough to make an eel mature, or whether it needs high pressure as well. But the eel is a slippery customer in more ways then one, and will probably keep biologists busy for many years to come.

A GREAT SOVIET BIOLOGIST

Most of the prominent scientists of the Soviet Union are over forty. So they were at least partly educated under Tsarism, and many had also studied abroad. The younger men and women, who hardly remember Tsarism and take Socialism for granted, have seldom done enough to achieve international reputation. But Lysenko is an exception. He is the son of a peasant, and only thirty-nine years old; and so far as I know, his first work was published in 1928 in the Caucasian Republic of Azerbaijan.

He has played a great part in the improvement of Soviet agriculture, and although some biologists doubt his theories, there is no question that his practical methods work. One of the main lines of crop improvement in the U.S.S.R. has been the selection of the best races of plant and the production of new ones by crossing. Here Vaviloff did great things for his country.

And, of course, the development of the electrical and mineral industries has assured a supply of chemical fertilisers. But Lysenko has studied, not so much the plant or its environment, but the relation between the two. Here is the problem. Everyone knows that seeds do not always germinate the moment they are sown. For, of course, wild plants generally sow their seeds in autumn, while they come up in spring.

Again, annual plants flower in their first year of growth, whilst others wait much longer. Thus if you try to grow a tulip from seed, instead of from a bulb, you will have to wait for at least seven years. Sometimes we want a plant to flower in its first year. For example, maize often fails to do so in England, and is killed by frost before it is any use. Sometimes we do not want it to flower. For example mangolds which flower in their first year use up the material stored in their roots to make flowers and seed. This is called "bolting," and if you look at a field of mangolds in early autumn you will generally see a few bolters.

Now Lysenko distinguishes sharply between development and growth. The seedling does not grow during the winter. It looks no different in March to what it did in October. But it has undergone an internal development which enables it to flower at the proper time. Similarly, all mangold and maize plants grow. But only some undergo the internal development which is needed to make them flower.

A very great deal of work has been done on the conditions for plant growth. But Lysenko was the pioneer in working out the conditions for development in seeds, while Garner and Allard in America did the same for the later stages in development. Their work has been extended by Lysenko and other Soviet biologists, such as Razumov and Liubimenko.

Lysenko's main practical problem was this. The summer is often so hot and dry in the Ukraine that the wheat plants may be damaged or killed unless they form their ears before the end of June. So rapid development is essential. Now, in England all kinds of wheat can be sown safely in the autumn, and the seedlings are not damaged by the winter frosts.

But this is not so in Canada and Russia, where the winters are very cold. The hardy but generally slow-growing forms which can be sown in autumn are called "winter wheats." The more delicate forms, which must be sown in spring, in some places as late as May, are called "spring wheats." If a winter wheat is sown in spring it may not produce ears at all, or may do so very late in the season. So many kinds of wheat which are useful in other countries are no good in the Ukraine.

Lysenko set out to treat the wheat seeds, before sowing, so that they could be sown in April, and yet get off the mark with a flying start, so to speak, and flower in June. The method, which is called *yarovizatzia*, or vernalisation (from the Latin *ver*, spring), differs for different wheats, but is as follows for some varieties:

The wheat is watered and kept at about 50° F. for twenty-four hours until a few seeds begin to sprout. Then it is spread out about 6 inches deep on the granary floor and the door and windows opened at night till the temperature falls to about 1° above freezing-point. The granary is shut in the day-time to keep it cool, and the seed stirred every day for a fortnight to a month, when it is ready for sowing.

As an example, one kind of wheat from Azerbaijan, if sown in the ordinary way, formed ears so late when sown at Odessa that it only gave 8 per cent. of the yield of a local wheat. When vernalised it ripened three weeks earlier, and gave a yield of 41 per cent. above the local variety. Of course, this technique, which requires a thermometer, good ventilation and careful weighing and measuring, is beyond the resources of an individual peasant, but quite easy on a big collective farm.

By 1937, 22 million acres were sown with vernalised crops. For the method does not apply to wheat alone. Maize must be kept in darkness for a fortnight at about 70° F. before sowing. In the case of potatoes, the aim is not, of course, to encourage seed formation, but that of tubers, and the treatment is quite different.

Instead of keeping them in the dark, they are threaded on string, and hung up in a greenhouse at 60–70° F., exposed to sunlight by day, and electric light at night. The idea of continuous lighting was due to Garner and Allard, but Lysenko and Dolgushin showed that it could be applied to the seed potatoes in an economical way.

Of course, we are only at the beginning of an understanding of the change which occurs in seeds and potatoes during vernalisation. These are being studied by biochemists in the Soviet Union. When they are worked out, still greater improvements will be possible. Lysenko is also tackling problems in plant breeding. Here he is engaged in a lively controversy with some of the older workers. This has not been very fully reported in English, but I hope that he may be able to attend the International Genetical Congress at Edinburgh next August to describe his work.[1]

Lysenko is not only an Academician, but a Deputy in the Supreme Soviet. He believes in a flying start for boys and girls as well as wheat and potatoes. "In our Soviet Union," he said "people are not born. Organisms are born, but people are made here—tractor-drivers, motor mechanics, academicians, scientists. I am one of these men who were made, not born. And one feels more than happy to be in such an environment."

[1] Unfortunately this was not possible.

BAD AIR

BAD AIR

A HUNDRED years ago all kinds of diseases were put down to bad air, and all sorts of futile precautions were taken against it. Great numbers of prisoners died of what was called "jail fever," and was actually typhus fever, which is carried by lice. To protect them against infection from the prisoners, the judges were given a bunch of flowers, which was supposed to purify the air. These flowers are still provided in some courts, just as the judges and lawyers still wear wigs, when everyone else has given them up.

This is a good example, not only of our out-of-date legal system, but of its tendency to treat symptoms rather than causes. Because unwashed prisoners smelt bad, the judges thought that they could protect themselves from their infection by a bunch of flowers which smelt good. Their attitude to crime to-day is not much more scientific.

How much truth is there in this view? How much illness and death is really due to bad air? Air may be dangerous to health for four different reasons. It may contain poisonous gases or vapours, harmful dust or disease germs. Or it may be unhealthy because it is hot and wet.

The largest body of men exposed to dangerous gases are the miners and workers in other similar trades, such as well-sinkers. In a well-ventilated pit, the gas is not dangerous unless there is a fire or an explosion. But where ventilation is bad several kinds of dangerous gas are met with. Fire-damp, or methane, is lighter than air, forms an explosive mixture with it and is also unsafe to breathe.

I was taught these facts in a practical way when I was about fifteen years old. I went down an abandoned coalpit near Cannock with my father and Professor (now Lord) Cadman. We crawled out of the main road into some old workings, and tested for fire-damp with our safety lamps. When we got to a place where we could stand up, there was so much gas near the roof that a safety lamp raised into it filled with blue flame and went out. If the flame had not been enclosed in wire gauze it would have set the gas alight and we should have been killed.

Then my father told me to stand up and recite poetry. I began on a bit of Shakespeare, but I tumbled down after two or three lines. This was safe enough, because the air at ground level was breathable. However, a commoner variety of bad air found in mines, and often called "black-damp," is heavier than air, so if a man is overcome by it he probably dies.

Black-damp (which of course is not black, but as transparent as ordinary air) is formed when coal or other easily oxidisable substances use up the oxygen of the air and replace it wholly or partly by carbon dioxide. Apart from coal-mines, it is found in wells, particularly when the barometer is falling. When this happens, the air in the pores of the soil expands as a result of the falling pressure and may fill a well or any other unventilated shaft.

So a belt of low pressure over England means not only rain, but danger to well-sinkers. Fortunately, where there is no danger of inflammable gas a candle can be used to test the air. It goes out long before the oxygen falls to a level which is dangerous to men.

The danger is more serious where compressed air is being used in the neighbourhood. A few years ago some shafts were being sunk in wet ground at Dagenham for the foundations of buildings. Some were so deep that compressed air was used to keep the water out. This air leaked through the mud, and in so doing lost all its oxygen. It came up through the shallower excavations, where compressed air was not needed to keep the water out, and killed five men.

Sewermen are occasionally overcome by hydrogen sulphide, a gas produced by putrefying materials, with a smell of rotten eggs. However, sewermen are accustomed to this smell, and to worse ones. And small quantities of this gas are not dangerous, so they are apt to disregard it.

The poisonous gas which kills most people in peace-time is carbon monoxide, which kills by uniting with the hæmoglobin of the blood in a way which I explained in an earlier article.[1] It is this gas which poisons coal-miners during a fire or after an explosion. And it is also the cause of death in coal-gas poisoning.

[1] See p. 190.

I have just been re-reading a report on coal-gas poisoning which was presented to the Home Secretary in 1899. My father, who wrote part of it, wished to determine how quickly gas would accumulate in an ordinary room if an unlit gas jet was left on. So he stayed in a room with gas escaping, making analyses, with a friend outside the window to pull him out if he fell over.

Rather to his surprise, he found that gas leaked out of most rooms very quickly, and he was never overcome by it. He did not even succeed in poisoning himself when he plugged up the crack under the door and other visible chinks. However, coal gas may sometimes accumulate in a room in dangerous amounts, even from a small leak, particularly if there is no wind and no great difference of temperature between the air inside and outside the house.

The leakage of outside air into closed rooms was first investigated by Peltenkoffer in Munich in 1858, and later work has only confirmed his findings. Here is a quotation from the Home Office Report of 1899: "He showed that the exchange of air is chiefly through the walls, and that the exchange of air was only diminished by 28 per cent. when all the chinks and cracks were carefully stopped up, and only increased to double when 8 square feet of window were opened, with a small temperature difference."

But in a strange document issued by the Home Office in 1938, called "The Protection of Your Home against Air Raids," we are told how to prepare a refuge-room against the entry of gas. "Fill in all cracks and crevices with putty or a pulp made of sodden newspaper. Paste paper over any cracks in the walls or ceiling." And so on.

I am not one of those who believe that the main danger in an air raid is from poison gas. Nevertheless, I regard these instructions as almost completely useless, and likely to divert the public attention from the real danger in air raids—namely, high explosive bombs. It would be interesting to know whether the officials who drew up the manual on air-raid protection had read the report for 1899, and if not, why not.

Besides the gases which I have mentioned, poisonous gases and vapours are used in many industries. There has been a good deal of illness from this cause in the rayon industry. We shall not know whether many workers have died until the next report of the Registrar-General on occupational mortality. These reports are supposed to come out every ten years, and the last was published in 1927. So the next is overdue.[1] Perhaps this is because it contains some facts which could be used by "unscrupulus agitators," who take too much interest in the health of the workers.

A particularly deadly vapour is that of lead tetra-ethyl, which is used as an anti-knock in petrol. A few years ago it killed about forty men in the United States, and others went mad from brain injury. We are told that the workers are now completely protected against it, but as lead may accumulate in the human body over long periods, this may only mean that they are not now being killed so quickly. Liquids whose vapours are poisonous are continually being introduced into industry as solvents, and unless the workers raise the matter, they are often left

[1] It was published in October, 1938, and I have written a series of articles on it (see pp. 165–178).

unprotected until a number have been killed, and many more have had their health ruined.

In later articles, I will deal with poisonous dusts and air-borne diseases.

BAD AIR IN THE HOME, SCHOOL AND BARRACKS

A CENTURY ago night air was considered dangerous. Nowadays many people think it unhealthy to sleep with the windows shut. I am writing about England, for I don't suppose many people keep their bedroom windows wide open on winter nights in Winnipeg or even Chicago. What scientific basis is there for this idea?

It is no good relying on general impressions. And statistics on matters of health may be very misleading. For example, I have little doubt that in England the people who use tooth-paste have better teeth than those who don't. But as those who can afford tooth-paste can also afford better food and better dentistry than the rest, this does not prove that the tooth-paste is the determining factor.

In the same way, people who live in overcrowded houses generally have inferior diets and no chance of taking regular baths. So it is hard to pin any particular kind of ill-health down to overcrowding. Fortunately, however, there are facts concerning men who had plenty of food and baths, but where overcrowding was found to cause disease, and there was no doubt that the disease was carried by air.

The most conclusive case is provided by the study of epidemics of cerebrospinal meningitis, or spotted fever, among soldiers in the Great War. This is a very deadly disease. About 60 per cent. of ordinary cases die, and this proportion can only be cut down to about 20 per cent. by serum treatment. However, it is rare. There were only about 2,000 cases among civilians in England and Wales during the epidemic of 1915.

The meningococcus, which causes it, is fairly common. In healthy populations, more than one person in 100 may harbour it in their throats. And during an epidemic it is found in anything up to half the men tested. In a small proportion of infected people, it penetrates into the fluid round the brain, and causes disease and usually death.

During the War barracks and huts were terribly overcrowded. After the Crimean War, Florence Nightingale had managed to get a standard laid down, which allowed each man 60 square feet of floor space, 600 cubic feet of air, and an interval of 3 feet between beds. On mobilisation half as many men again were crowded in, and finally the number was doubled, so that beds were less than 6 inches apart. Finally, on account of anti-aircraft lighting restrictions, windows were covered over with blankets or blackened, and shut at night even in summer.

Under these conditions, the meningococcus spreads rapidly. With beds 3 feet apart less than 2 per cent. of men harbour the meningococcus. With an interval of 2 feet it rises to about 8 per cent. and at 1 foot to 20 per cent. But how do we know that infection takes place mainly in sleeping quarters and not in living-rooms? For one thing, people in neighbouring beds were commonly infected. For another, overcrowding in working places does not cause meningitis, though it favours the spread of influenza.

One reason for the spread during sleep is that many people sleep with their mouths open. During the day the little drops of water carrying disease germs are largely stopped in the nose before they get down to the throat, which is far more vulnerable. Hence epidemics during the war were wholly due to overcrowding in sleeping quarters. Dr. Glover studied four epidemics in the Guards' Depot at Caterham, and was able to predict the onset of the last three.

A similar study was made by Surgeon-Commander Dudley on the boys of the Royal Naval School at Greenwich, bearing on the spread of diphtheria and scarlet fever. Here the beds were only $1\frac{3}{4}$ feet apart, though, as the room was high, each boy had over 500 cubic feet of air. Here, again, a map of the dormitory made it quite clear that the infection spread from bed to bed. In one dormitory with 126 boys there were nineteen cases of diphtheria, eight of which were in one block of twelve beds. The spread from one dormitory to another was put down to sucking pencils and penholders. Virulent diphtheria bacilli were found on a penholder which had been put away for fourteen days. The day-boys, who brought their own pens in boxes, were never infected.

Apart from infection of this sort, experience seems to show that diphtheria rarely spreads when beds are 10 feet apart. Of course, the infection is greatly lessened when windows are open, and this is one reason why epidemics generally occur in cold weather or about three weeks after cold weather. Apart from the fact that germs are blown away when windows are open, they survive very much longer in moist than in dry air.

The moral seems to be that the main hygienic advantage of open bedroom windows is in overcrowded bedrooms. If everyone in Britain except married couples and babies had a bedroom to themselves we should be very much healthier, and there would be little or no need to keep windows open in cold weather. The main exception to this rule is pulmonary tuberculosis, where cold air seems to help you to kill your own germs, as well as protecting you from other people's.

If, in addition, food and milk were more hygienically prepared, and crockery, glasses, spoons and so on in public eating and drinking places were adequately sterilised, it is probable that a large group of diseases, including scarlet fever and diphtheria, could be made extremely rare, if not wiped out. The isolation of those who are actually ill will never be sufficient, because healthy people or, at any rate, people with mildly sore throats, can carry the infection.

How many more diseases are spread in the same kind of way we do not know. Rheumatic fever, which is one of the main causes of heart disease, is certainly associated with overcrowding to such a degree that it is hard to doubt that it falls into the same category.

Other air-borne diseases, such as influenza, are much more infectious, and there is little doubt that they can spread in crowded workshops, offices, schools and trains, where diphtheria does not spread. We are only at the beginning of a scientific attack on air-borne diseases which will lead to a scientific standard of housing as definite as our standards for drinking water and diet.

Meanwhile our ventilation standards are rather out of date. The carbon dioxide in the air of schools and factories must be kept below a certain level. Carbon dioxide gives a measure of the pollution of the air by breathing. But it is not harmful in itself until it rises to a level far above that legally allowed.

It is only an indicator of danger. The same is true of bad smells. They are very rarely directly harmful, or sewermen and tanners would be very unhealthy, which they are not. They are danger signals of decaying matter which may breed flies or pollute food. And air with no bad smell or an excessive amount of carbon dioxide may be dangerously infectious. We have no scientific standard of overcrowding, but what has been learned about soldiers and sailors makes it sure that, when such a standard is discovered, a large proportion of our civilian population will be found very far below it.

BAD AIR IN THE FACTORY

In the factory men are exposed to unnatural conditions. Our senses generally, though not always, give us warning of the kinds of bad air which we are likely to meet in Nature. They smell bad, like the air near rotten meat, or sting our eyes, like the smoke from a fire. But many of the liquids used in industry have quite a pleasant smell, though their vapours are poisonous.

Thus tetrachlorethane is used as a solvent for cellulose acetate in many industries, including rayon, aeroplane "doping," the boot trade, and even the manufacture of artificial pearls. It has a smell like camphor, which some people do not find unpleasant. In strong concentrations it irritates the throat. But when breathed for long periods it may affect the nervous system, causing trembling and paralysis. Or it may cause death from liver disease.

Carbon disulphide, which is used in the rubber, rayon and lacquer industries, is fortunately generally so impure that it has a bad smell, which gives some warning, though when pure the smell is not unpleasant. But even when pure it is a general poison for the nervous system. In the German rubber trade, madness was found to be seven times as common among women workers as in the general population.

The vast majority of solvents have more or less poisonous vapours, and the workers are generally protected to some extent against acute poisoning. It is very difficult to prove that a case of liver or brain disease is due to industrial poisoning, and harder still to get compensation. And since the use of solvents has enormously increased since 1918, we shall know very little about their killing power until the publication of the next report on occupational mortality by the Registrar-General.[1]

The earlier reports, however, give a lot of information

[1] We still know very little about it, as the trades were not so classified as to bring out the facts.

as to the terrible effects of dust in various industries. In 1921-3 the death-rates in 164 occupations from a number of different causes were determined. The seven occupations with the highest death rates from all causes together were cutlery grinders, tin-miners, barmen, file-cutters, kiln and oven men in the china and earthenware trades, street hawkers and potters.

The seven occupations with the highest death-rate from respiratory diseases were cutlery grinders, tin-miners, china and earthenware kiln and oven men, cotton strippers and grinders, potters, cotton blow-room operatives and stevedores. So it is clear that the main cause of death in these industries must be from the air.

This is largely true even for the barmen, who had nearly twice the normal death-rate from respiratory diseases, but who, of course, also die from digestive and other diseases due to overdrinking. It is to be noted that these death-rates include the rates from accident. These are, of course, terribly high among coal-miners, railway shunters and bargemen. But accidents do not take such a high toll of life as respiratory diseases, even among coal-miners.

Let us look at some of these trades in detail. Tin-miners die of silicosis, but, as tin-mining in England is a dying industry, this is relatively unimportant. The cutlery grinders, mainly in Sheffield, died mainly of bronchitis and phthisis from the silica dust caused by the abrasive used, in this case natural sandstone. But they also had the highest mortality of any occupation from influenza, and the highest but one both from cancer, pneumonia and one type of heart disease.

The other metal-grinders who had a death-rate only just below that of potters mostly used abrasives such as carborundum and emery, and many used a wet-grinding process, which produces relatively little dust. Even so there can be little doubt that the large increase in the number of metal-grinders in the motor and aeroplane trades in recent years will cause many deaths. But as the poisoning by these dusts often takes ten to twenty years to kill a man, the full effects are probably not yet visible.

The potters had over five times the normal death-rate from bronchitis, and the kiln and oven men in the pottery trade nearly five times. This is certainly due to the dust inhaled, for brick and tile kiln and oven workers are healthier than the average of the population. There is no doubt that bronchitis should be classed as an occupational disease in the pottery trade, as silicosis is in many others.

There are fortunately not many file-cutters. Their deaths from lung disease are high, whilst they head the list for deaths from kidney disease, and come very near the top with apoplexy and some forms of heart disease. These are probably due to breathing lead dust, though no deaths were registered as due to lead poisoning. However, lead probably kills in different ways, according as it is breathed or swallowed.

All workers in the cotton carding-rooms and blow-rooms are subject to heavy mortality from dust. Here again the inhaled dust affects not only the lungs, but kidneys and several other organs. There can be little doubt that this mortality could be greatly lessened if it received greater publicity. Stevedores inhale a great deal of dust, and are a little less healthy than the other dock labourers. But all dockers are very unhealthy.

It must not be assumed that all dusts are dangerous. Coal-miners had less phthisis than the average,[1] though a good deal more bronchitis, especially in old age. But cement workers and lime-burners, who inhale plenty of lime dust, are conspicuously healthy, and had fewer deaths from lung diseases than the general population. And limestone miners and quarry-men were healthier than the average, whereas sandstone workers had a very high death-rate. Again, grain-millers, though notoriously dusty, had less deaths from lung disease than the average.

It is up to the trade unions concerned to do what they can to check the death-rate, both from poisonous vapours and from dust. In many cases research is needed. But it is unlikely that research will be undertaken on a big scale unless the employers are compelled to give compensation for illness and death from industrial diseases. We do not know, for example, just why kidney disease is so prevalent in the dusty branches of the cotton and woollen trades. If the employers had to give compensation for it, we should probably know fairly soon.

In the nineteenth century there was an agitation against such scandals as lead and phosphorus poisoning which has made them relatively rare. The effects of dust are equally preventable. But they will not be prevented unless the workers take the matter in hand.

[1] But see p. 172 for more up-to-date figures.

COLLIERY EXPLOSIONS

THE terrible disaster at Markham[1] has made colliery explosions news once more. We shall not learn the full truth about what happened for some time, if at all. For the men who saw the explosion start are probably dead, and the Gresford enquiry showed that, if regulations were broken, great efforts will be made to conceal the facts.

What is an explosion? The answer may, alas, concern all of us in the near future, as it concerns the people of Barcelona and Valencia to-day.[1] An explosion is a sudden change by which a large volume of gas is generated. The change may be a physical change, as when a cylinder of compressed air bursts. But it is almost always a chemical change which produces a great deal of heat.

The heat alone is not enough. When an incendiary bomb containing thermite is lit, it lets out a flood of white-hot molten iron. It may set a house alight, but it does not wreck even a single room, as it would do if the same amount of heat were used to make gas expand suddenly.

Most explosives are solid. When gunpowder or gelignite explodes, the solid is turned into very hot and intensely compressed gas, which will expand till it occupies many thousand times the volume of the original solid, and pushes away anything which prevents it from expanding. Liquid explosives, such as nitroglycerine, behave in just the same way.

[1] May, 1938.

In a colliery explosion the explosive is either gas, a mixture of air and fire-damp; or partly gas and partly solid, a mixture of air and coal-dust. Fire-damp, or methane, has been accumulating in the coal for millions of years, and comes out when it is hewn. It is not poisonous unless there is about 50 per cent. of it mixed with air, in which case the amount of oxygen becomes dangerously low.

But when anything between 6 per cent. and 13 per cent. of fire-damp is mixed with air, the mixture is explosive. A flame once started travels quickly through the mixture, gathering speed. The heat causes the air to expand, and great damage is caused, especially when props are blown out, and the roof falls.

However, big gas explosions are now rare, for two reasons. Davy invented the safety lamp, in which the flame is enclosed in glass, and the air going to and from it has to pass through wire gauze. Even if the air inside the glass chimney explodes, the flame is stopped by the gauze, which cools the hot gas passing through it.

Besides this, if a mine is properly ventilated the gas never reaches an explosive concentration. It can be detected in amounts less than 1 per cent. For if the wick of the safety lamp is turned down till there is only a small blue flame, the gas can be seen burning above it. In a properly ventilated mine an explosive concentration of gas can only occur as the result of a sudden outburst.

For a long time mining engineers did not believe in dust explosions. Even a thick cloud of coal-dust is hard to light. An ordinary flame will not set it off. But it can be lit by a small gas explosion or a blown-out shot from blasting, though the danger is lessened when "permitted" explosives are used.

Until 1908 many mining engineers did not believe in dust explosions. Then the Mining Association, which was a much more progressive body than it is now, built a gallery at Altofts Colliery, Normanton, Yorkshire. This was made of old boilers about 7 feet in diameter and placed end to end. Coal-dust was placed in shelves along it, and two cannon were put inside.

One contained about 1 lb. of gunpowder, and was let off to stir the dust up. A few seconds later the larger one, with 10 lb. of powder, set it alight, and the mixture exploded. In the first few experiments the explosion was only allowed to travel for about 100 feet, and was very mild. In fact, some people still thought that dust could not account for the damage done in explosions underground.

I was present at the experiment which decided the matter. The gallery had been lengthened for another 100 feet or so, and the explosion was given a longer run than ever before. I was about 400 yards away; and I saw a huge cloud of smoke, and what looked like pieces of burnt blackened paper sailing into the air. These were bits of the boilers, fantastically twisted, and one flew over our heads with a sound which I did not hear again till I was shelled in 1915.

As the path of the explosion lengthened there had been a change of quantity into quality. From being a mild puff which did not break the boilers it had become a shock so violent that, although the news was kept out of the papers (the Mining Association can manage that kind of thing) it shook houses six miles away, and was reported as an earthquake!

This sudden change is well known in gas explosions, but there it takes place after a run of a few inches. As a piece of boiler as big as an ordinary table had been flung over the main Midland Railway line the experiments were moved to a Derbyshire moor. And there it was found, largely through the efforts of Sir William Garforth, who also designed one of the first rescue apparatus, that if an equal amount of stone dust is mixed with coal-dust, the mixture will not explode.

The earlier method of watering was not only difficult and costly, but was liable to cause falls of roof. So stone dusting was introduced, and besides this various precautions to prevent the accumulation of dust in the main roads were made compulsory. Where these precautions are properly carried out, large dust explosions are impossible.

Of recent years, however, the danger of explosions has been increased by the introduction of electrically driven coal-cutting machinery. A spark from such a machine caused the explosion at Markham in January, 1937.

Most of the deaths in colliery explosions are not due to violence, but to poisoning by after-damp, as described on page 190, or to burns. My late father worked on both these subjects. Along with Dr. Wade and others, he was responsible for introducing tannic acid as a treatment for burns. His arms were covered with the scars of experimental burns on which he had tried different forms of treatment.

As early as 1896, in a report to the Home Secretary,

my father recommended that men working at the face should be provided with oxygen apparatus with which they could make their way through poisonous air to the shaft. In 1915 he recommended that soldiers should be given box respirators. This was at first thought unnecessary, but it was done in 1916. However, no one took the recommendation of 1896 seriously.

He had hoped that, with better ventilation and stone dusting, the provision of rescue apparatus for all miners would no longer be necessary. But to-day, owing to neglect of regulations and the growing use of electrical machinery, big explosions seem to be getting commoner. So perhaps it is time that the miners, who risk their lives for the rest of us every day, should be protected against poisonous gases as the soldiers are protected.

COMPRESSED AIR ILLNESS

A CASE of compressed air illness[1] has recently been reported in a worker on the new Purfleet-Dartford tunnel. And for one case of this complaint that gets into the Press, dozens are kept quiet. So, though it is not one of the more serious industrial diseases, it is one of which everyone interested in workers' health should know something.

When a tunnel is being bored in wet ground, water and mud tend to pour into it. That is why the deeper railways under London have to run in watertight steel tubes. But while they are being excavated one end of the tube is

[1] Since this article was printed there has been a fatal case.

open. To prevent water coming into this, compressed air is pumped in.[1] To balance a column of 34 feet of fresh water or 33 feet of sea water an air pressure of 1 atmosphere, or 15 lb. per square inch, is needed.

In the same way a diver in an ordinary flexible dress has to breathe compressed air. For example, at 30 fathoms he is at a pressure of 6 atmospheres extra. In the modern armour-plated dresses this is not necessary. But they are clumsy and expensive, and the diver cannot use his hands, though he is probably safer than in the old type.

Now, pressure by itself is not dangerous provided that it is even. Workers who dive for pearls without a dress are often killed by the pressure of the water on their chests. But that is because they have no compressed air in their lungs to balance it. In the same way a diver with a bad cold may burst his ear drum from unbalanced pressure, because the tube leading from his throat to the inside of his ear drum is blocked.

The danger comes, not when the pressure is put on, but when it is taken off. Among the effects of this may be violent pain, paralysis, squinting, fainting, and sudden death. It is also very likely that several kinds of chronic illness may be caused by compressed air, but not so likely that compensation will be given for them.

The cause of the illness is as follows. When any liquid is exposed to a gas, some of the gas dissolves in it. Ordinary water contains some air in solution. Cold water dissolves more than hot, so when water is boiled the air appears as small bubbles before the water itself boils.

[1] This is only needed in those parts of London where the soil is very wet.

The amount of gas dissolved is proportional to the pressure. Soda-water is made by forcing carbon dioxide under pressure into water. When the pressure is lowered, the water cannot hold so much gas as before, and the extra gas comes off in bubbles.

Now, all the time that a man is breathing compressed air, more air than usual goes into his blood at each breath, not, of course, in the form of bubbles, but dissolved and therefore invisible. However, it can easily be made visible by taking a sample of blood from a vein, and then extracting the gas from it with a pump.

Air consists of oxygen and nitrogen. The extra oxygen is used up, but the nitrogen goes into the organs of the body, particularly into fat and into the nervous system, which contains a lot of greasy substances in which nitrogen is very soluble.

When one of the first tube railways was made under the Thames, the directors of the company were invited to a luncheon when the tunnel got to the deepest point. They lunched in the compressed-air chamber, and were, of course, given champagne. They complained bitterly that it did not fizz. This was to be expected, because the wine was still under pressure when the bottles were uncorked. However, they drank it. After lunch they went into the air-lock and the pressure was lowered. The champagne began to bubble, and the directors knew all about it.

Now bubbles in the stomach are only a nuisance. But in the nerves or blood stream they are a danger. If a man who has been breathing air at 2 or 3 atmospheres excess pressure for several hours comes back to the normal pressure of 1 atmosphere too quickly, bubbles of nitrogen are almost sure to form somewhere in his body. One of the commonest places is the bag of fluid between the bones at a joint, where they cause pains like those of violent rheumatism, called "bends." If they form in the nervous system, they cause pain or paralysis. If in the blood they may cause death. For the frothy blood accumulates in the lungs, and cannot be forced through the small blood vessels. The victim struggles for breath, goes black in the face, and dies of suffocation. Fortunately, pain or paralysis, which seldom last more than a few hours, are much commoner than death.

The obvious way to prevent these symptoms is for the diver to come up very slowly, or for the men who have been working in compressed air to stay a long time in an air-lock, where the pressure is very gradually lowered. But this is too simple. For some parts of the body, such as joints, take up nitrogen very slowly and give it out slowly, too. And during the first part of the decompression, while the pressure is still high, these parts will actually be taking up more gas, and the danger will be increasing.

The problem was first scientifically tackled about thirty years ago by my late father, Commander Damant, R.N., and Professor Boycott, for the British Navy. They found that a diver could always come up a little over halfway to the surface without danger, and that he could then ascend safely by stages which were worked out according to theory and thoroughly tested.

However, the tables of decompression which they drew up for divers have been modified for compressed-air work, and the fact that men occasionally fall ill shows that they are sometimes decompressed too quickly. The fact that tabs are sewn to their clothes, or badges are worn, giving instructions as to what should be done if the men fall unconscious, is sufficient testimony to the danger of the work.

It is to be hoped that the trade union officials concerned have made themselves familiar with the existing rules for decompression, and will not only see that they are rigidly adhered to, but that if in spite of them accidents take place, the later stages of decompression are still further slowed down. It is only by constant vigilance that the workers can ensure their safety; and this vigilance must be based on technical knowledge.

THE STRANGE CASE OF RAHMAN BEY

IN several daily newspapers of July 1st, 1938, there was a story of an Egyptian called Rahman Bey, who threw himself into a trance, and stayed for an hour at the bottom of a swimming bath in a metal tank. "My peculiar gift was discovered by a Yogi priest when I was a child," he is reported to have said. And the whole thing is put across to the British public as a sample of the mysterious gifts of Orientals, who, of course, are so unlike us!

Now there are two funny things about this story, before we come to the tank at all. The name Rahman Bey means Colonel Merciful, which strikes me as funny. And as the Yogi philosophy is a product of India, it is remarkable that its adherents should experiment on Egyptian children.

The *Daily Herald* gave a photograph of the tank, which measures 8 feet by 1½ feet by 2 feet, according to one of my colleagues, and rather less according to me. So it holds 20 to 25 cubic feet. An average man occupies 2½ cubic feet, so Rahman Bey had about 20 cubic feet of air. Now a man doing light work uses about 24 cubic feet of oxygen in 24 hours. If he lies still, this is reduced to about half.

So Colonel Merciful used about half a cubic foot of oxygen in an hour. As 20 cubic feet of air contain just over 4 cubic feet of oxygen, he had plenty to spare at the end of the hour, and could have gone on for another two hours. After this time he would have been very short of breath, and would have panted so much that the remaining oxygen would have been used up more quickly. And when he came out after three hours he would have had a nasty headache. For besides using oxygen, a man makes a slightly smaller amount of carbon dioxide, and after breathing air containing anything over 6 per cent. of this gas for an hour, one has a short but violent headache, and I for one sometimes vomit.[1]

Before being shut up in the tank, Rahman Bey "shook like a pneumatic drill and then flung himself violently into unconsciousness." I should have lain down quietly. But if Rahman Bey was unconscious he saved himself an hour of boredom. However, if I can borrow a tank, I am perfectly willing to spend an hour at the bottom of a swimming bath for a suitable fee to be paid to the International Brigade Dependants Fund.[2]

Many readers will say "What does all this matter?" It

[1] So do others, as the experiments on the International Brigaders in connection with the loss of the *Thetis* showed.

[2] Unfortunately, this challenge was not taken up.

matters quite a lot. The physiology of human breathing is involved in questions of mine and factory ventilation, and of protection against poison gas, which to-day concerns everybody.[1] And because we are not educated in the matter, we take the Government's statements on gas defence as seriously as the journalists took Rahman Bey.

I recently had an airtight tank with a glass window made in which a child could be shut up for several hours during a gas attack. I stayed in it for an hour myself, and got rather warm, whereas Rahman Bey was doubtless cool at the bottom of his swimming bath. But I had to try three mothers before any of them would allow their baby to stay in it for even half an hour.

There are no gas masks for babies,[2] and a tank of this kind would give protection for some hours at any rate. But because we are not taught such elementary facts about ourselves as how much oxygen a man uses per hour, a good many babies are going to die if gas is used against civilians in a future war.

You may call me a crass materialist, but it seems to me more important that children should be taught such facts as this than that they should know how often King Henry VIII married or who won the Battle of Agincourt.

There are a great many strange stories about the wonderful powers displayed by various Asiatics and Africans. When they are investigated they generally turn out either to be untrue or to be based on elementary facts of human physiology which are known to certain groups in India, but not yet generally known in Europe.

[1] As the *Thetis* case showed, it is also important to submarine crews.
[2] A few are now available.

For example, I have pushed a red-hot cigarette end against the finger of a hypnotised Englishman without causing either pain or blistering. If he had been an Indian, it would have made a story for the daily Press.

These stories are very useful to imperialists, because they help to spread the idea that the human races are very different. If people in England believe the myth that members of coloured races have powers which Europeans do not possess, they will be ready to believe another myth —namely, that they do not possess the power of looking after their own affairs.

It is time that we realised that scientific investigation has shown that people of different races are remarkably alike, and that it is only prejudice and the self-interest of exploiters which prevent them from being brothers.

MEDICINE AND SOCIETY

OCCUPATIONAL MORTALITY

WHICH is the most dangerous trade? Which is the healthiest? How far is the ill-health in an industry due to bad working conditions or to bad housing and low wages? Which diseases attack the poor and which the rich most particularly? What diseases should be regarded as industrial diseases for purposes of compensation?

These are some of the questions which can be answered from the Registrar-General's Report on Occupational Mortality, recently[1] published. It is a document in the finest tradition of our Civil Service, the tradition of the reports of factory inspectors which Marx found so valuable when he was writing *Capital.* And nothing comparable is published in any other country. So we have to judge of conditions abroad from our own. Parts of it show real improvements in health. But it also tells some terrible stories of lives wantonly thrown away in many industries, and of poverty killing men, women and children.

I am going to devote several articles to it, so I must first describe how the figures were obtained. The numbers of men in each trade, and of women in some, were obtained in the 1931 Census and compared with the numbers

[1] November, 1938.

dying in 1930, 1931 and 1932. Thus of 101,157 bricklayers between twenty and sixty-five years, 2,322 died. More than 300 died of cancer, of heart disease and of consumption; and smaller numbers from each of a number of other causes.

But this is not good enough as a measure of their death-rate, for the following reason. Out of 9,089 gamekeepers between the same ages, 190 died, which is almost the same fraction as for the bricklayers. But they were really a great deal healthier. For there were more bricklayers between the ages of twenty-five and thirty-five than between fifty-five and sixty-five, and naturally enough the younger men are healthier. But gamekeeping is a dying occupation, and there are more gamekeepers between fifty-five and sixty-five than between twenty-five and thirty-five.

So until allowance is made for this fact gamekeepers appear to be less long-lived than is really the case. The difficulty is got over as follows. The statisticians calculate how many bricklayers at each age would have died if they had had the same chances of death as the general population. They add up these figures, and by dividing them into the actual deaths obtain what is called the "standard mortality rate."

When this is used for comparing different industries, the results are clear. The seven deadliest occupations are tin and copper miners, sandblasters, metalliferous miners (other than tin, copper and iron miners), stevedores, glass blowers, slate miners, and kiln and oven men in the pottery trade. And in each case their high mortality is largely due to lung disease, mainly caused by the dust which they inhale.

The tin and copper miners head the list for deaths from respiratory tuberculosis (consumption) and chronic interstitial pneumonia (silicosis), while sandblasters have the highest figure, over seven times the normal, for respiratory diseases in general. Other particularly unhealthy occupations are innkeepers, glaziers, cotton blow-room operatives, pottery dippers, barmen and bookmakers.

It is not so clear what is the healthiest trade. The figures show the least deaths among warehouse and storekeepers' assistants, but this means very little, for the following reason. A man may state his occupation correctly at the census, but his widow or next-of-kin gives him too exalted an occupation when he dies. So men are registered as navvies at the census, but builders' labourers when they die, and navvies in the building trade appear to be very healthy.

When such allowances are made, the healthiest occupations appear to be agricultural machine workers, wireless operators on shore, lime-burners and kilnmen, draughtsmen and costing clerks, bank and insurance officials. For a long time Anglican clergymen headed the list, but they are now twelfth out of 200, though they still delay their departure for a better world longer than clergy of other denominations.

However, the very long-lived or short-lived occupations generally include a few thousand men only. It is more interesting to see which of the larger occupational groups are particularly healthy or otherwise. Most of the big groups—for example, miners, metal workers, and the distributive trades—have a mortality within 10 per cent. of the average.

There are seven large groups with a mortality less than 90 per cent.—namely, agricultural workers, Civil Servants, professional workers (including teachers), railwaymen, woodworkers, builders and printers. At the other end of the scale come men engaged in the distribution of alcohol (innkeepers, barmen and waiters), water transport workers including seamen and dockers, and general labourers and other unskilled workers.

The safest of the large occupations is agriculture, with a death-rate 27 per cent. below normal, and the most dangerous is the drink trade, with mortality 51 per cent. above it. But this only employs 110,000 men, compared with over 1,000,000 general labourers and other unskilled workers and nearly half a million water transport workers. It is these two groups that contribute most to the excessive deaths, and very largely on account of sheer poverty rather than special occupational risks.

INDUSTRIAL ACCIDENTS

WHEN we think of a dangerous trade, we often mean one with a great many accidents—for example, that of a railway platelayer. Platelayers certainly have a terrific death-rate from accidents, but their death-rate from all causes is very close to the average, because their mortality from phthisis is about two-thirds of the death-rate of men in general, since they get plenty of fresh air. And their gain on deaths from phthisis balances their loss from accidents.

So accidents hardly ever account for more than about

one-seventh of the total deaths, except in the case of airmen, for whom the accident figures are not given separately in the Registrar-General's Report on Occupational Mortality, but whose very high death-rate is no doubt largely due to accidents. If we except these, the most dangerous occupations are stevedores, slingers and riggers, slate-miners, coal-miners, seamen, stone-miners, shunters, pointsmen and level crossing men, bargemen and boatmen, and platelayers.

The figures for seamen are a little doubtful, as, of course, many of the registered deaths occurred abroad, and accidents were grouped together with other kinds of violent death. However, there can be no doubt about the stevedores and bargemen. It is very interesting to note that the fatal accidents among fishermen were hardly above the average, and their general death-rate slightly below it. The reason is probably that fishermen have a good deal more say in what risks they take than seamen on large ships, besides having a healthier life in other respects, getting a lot more fresh air.

All kinds of miners and quarrymen have a very high death-rate from accidents. If the mortality of coal-miners from accidents could be brought down to the average level, coal-mining would actually be a safer occupation than most, in spite of the high mortality of anthracite miners from silicosis.

One reason for the high death-rate among miners is doubtless the lack of publicity. In 1930–2 more than two coal-miners, on an average, were killed every weekday. But a miner's death is not news unless a number are killed together in an explosion. We hear hardly anything of fatal accidents to slate-miners and quarrymen, though in 1930–2 slate-mining was a little more dangerous than coal-mining from the point of view of accidents, and much more so from that of disease.

I believe that if the *Daily Worker* would publish the names of the miners accidentally killed every day, the rest of us might begin to realise to what risks the miners are exposed. Just the same is true in the dangerous branches of the transport industry. If a passenger is injured in a collision, that is a railway accident, and news. If a shunter or platelayer is killed (and more than one is killed each week), that is not news.

But the two most dangerous trades of all—stevedores and slingers and riggers—are concerned with cranes. I must confess that until I read the Registrar-General's report I did not realise that when it comes to killing workers, cranes are more efficient than mines or locomotives, and only excelled by aeroplanes. And I am sure that it is not generally realised in the Labour movement.

Two women's trades—waitresses and charwomen—had death-rates from accident nearly double the women's average. But, of course, for women as a whole childbirth takes the place as a cause of death that accident does among men, though the total deaths of women from this cause are less than half those of men from accident.

What should be done to bring the accident rate down? No doubt a good deal could be done by seeing that existing safety regulations are carried out. But this is not so easy in practice, and it might be better to provide an economic incentive. Employers and their representatives are apt to look at accidents from the economic angle.

In a recent discussion at the Institution of Mining Engineers, one speaker said that coal-mining was becoming safer, because the number of accidents per ton of coal cut with mechanical cutters was falling. It was pointed out to him that the accident rate per man-hour was going up—in other words, that a miner now stood a bigger chance in each shift of being killed or injured. But he did not think this mattered much.

If employers think in terms of money, not men, as in this case, they might take steps to reduce the accident rate if the compensation payable at death were raised until it became worth their while to do so. It should never be forgotten that compensation acts not only as a partial payment for past accidents, but as a preventive for future accidents, too.

No doubt until workers control industry, accidents will not be reduced as they could be. But that is not to say that we cannot do something to-day by seeing that existing regulations are kept, new regulations made, and an economic incentive given to employers to prevent accidents.

INDUSTRIAL DISEASES

IN 1897 British workers established the principle of compensation for death or injury while at work, and the scope of the original Act has been extended to cover some kinds of industrial disease, such as lead and phosphorus poisoning and, later on, tar cancer and silicosis. But even now the majority of industrial diseases are not compensated in any way.

The most serious occupational diseases, from the killing point of view, are those of the lungs. A small fraction of all deaths from lung disease are compensated as being due to silicosis, particularly in the two occupations—sand-blasters and tin and copper miners—whose total death-rate from lung diseases is over seven times the average.

But this is only done where the disease can be shown with absolute certainty to be due to the conditions of labour. Let us see how this works out. Deaths from silicosis are registered as due partly to "chronic interstitial pneumonia" and partly to "respiratory tuberculosis"—that is to say, consumption affecting the poisoned lungs. We can get some idea of the relative frequency of silicosis from the fact that four agricultural labourers out of 400,000 died of chronic interstitial pneumonia in 1930–2 as compared with ten out of 6,813 Welsh anthracite miners and eight out of 4,988 sandstone quarrymen. That is to say, the death-rate was about 150 times as great in the two dangerous occupations.

But the death-rate of these miners and quarrymen from lung diseases other than consumption and chronic interstitial pneumonia was two and a half times that of the agricultural workers, and wherever there is a high death-rate from silicosis we also find high death-rates from other lung diseases which are not compensated. We could almost make a geological map of England on the basis of the deaths from lung disease among quarrymen. Limestone quarrymen have a death-rate from all lung diseases of only 44 per cent. of the average. There can be little doubt that limestone dust is good for the lungs. Workers in igneous rocks have exactly the average death-rate from this cause. But workers in sandstone (Lancashire, Cheshire and the West Riding) have a death-rate 86 per cent. above the average, and North Welsh slate quarrymen 122 per cent. above. Sandstone masons, with a death-rate from lung diseases of 239 per cent. above the average, are one of the most unhealthy of all trades.

In each case only a small fraction of all deaths from lung disease is attributed directly to silicosis, and therefore ranks for compensation. It is not only mineral dust which kills. Cotton strippers and grinders have a death-rate over double the average from lung diseases, and cotton blow-room workers, though their death-rate from this cause has improved, are still being killed off by dust.

And it is not only the lungs that suffer in these dusty trades. Mineral dust does not seem to affect the kidneys, but textile dust does so. Although the numbers are small, cotton blow-room workers head the list of deaths from this cause, with three times the average, and cotton strippers and grinders, wool-weavers and wool-spinners are all very liable to it.

What could be done to lower these death-rates? I believe that the first step should be as follows: Whenever the death-rate from a particular disease in any industry is double the average or more, it should be registered as an industrial disease, since the presumption is that a death from it is due to the conditions in the industry. Thus twenty-two sandstone masons between the ages of twenty and sixty-five died of bronchitis in 1930–2. Only eight would have done so if they had had the average death-rate. The odds are seven to four that any given case was due to employment, and even if each widow only got seven-elevenths of the full compensation, the industry would pay for the deaths which it causes.

If compensation was given in such cases two things would happen. Employers would install proper ventilation and issue respirators where necessary. And they would demand research on the diseases. We have no idea why textile workers get kidney disease, nor why, to take another example, four out of the five trades with the highest death-rates from pernicious anæmia—namely, stationary engine drivers, railway engine drivers, coal-mine workers above ground and coal-heavers—are all connected with coal.

If the trade unions paid more attention to industrial diseases, they could probably induce the Medical Research Council to investigate such matters. Meanwhile, however, the majority of industrial diseases are not officially recognised as such, and employers fight against their recognition. The Labour movement should work to get the list extended, and to encourage investigations on their cause and cure.

This will involve not only agitation, but statistical research. The Labour movement is quite right, from the point of view of life-saving, to concentrate on wages, since poverty kills more people than do bad conditions of work. Nevertheless, industrial diseases are terrible killers, and every active trade unionist would do well to study the health of his or her own occupation, and to see how it can be improved.

THE ECONOMICS OF CANCER

CRITICS of Marxism often accuse us Marxists of dragging economics into every discussion. So it is worth while showing that you cannot possibly keep economics out of a discussion even of cancer. Cancer is a rather mysterious disease in which the cells in a small section of some organ, instead of merely replacing other cells which die, start to multiply, and, what is more, to invade other organs.

If cells merely increase in number instead of migrating, we get what is called a benign tumour, a mere lump which may be quite harmless. But once they start founding colonies in other organs, say little patches of stomach cells in the liver, lungs or lymph glands, death is bound to occur unless all the colonies can be destroyed, which is rarely possible.

Fortunately, cancer is fairly rare up to the age of forty-five. But from then onward it increases rapidly. As a result, the death-rate from cancer is going up. For if we eradicate a disease like typhoid fever, which kills the young quite as easily as the old, some people who would otherwise have died of typhoid will die of cancer. But there is no reason to think that the chance of dying of cancer at any particular age is increasing.

There is a hereditary factor in cancer. That is to say, if one of your parents had cancer you are rather more likely than the average person to develop it. But even if both your parents died of it, you can comfort yourself with the fact that even so, you will probably die of something else. The environment also has a considerable effect. Otherwise all occupations would have the same cancer death-rates, apart from small differences due to chance, when allowance has been made for age. This is not so.

It has long been known that a few trades had a very high mortality from some kinds of skin cancer. Chimney-sweeps and mule-spinners are particulary liable to cancer of the skin of the testicles, due to chronic irritation by soot and lubricating oil. At one time chimney-sweeps headed the list of cancer death-rates. This is no longer the case, mainly because they have been able to adopt higher standards of cleanliness. But their death-rate from cancer is still far above the average. Fifty years ago the chimney-sweep's black skin was regarded as a joke. We now realise that if a sweep has to live in a house without proper washing arrangements, he is probably condemned to death from a particularly unpleasant kind of cancer. Gas-workers are affected in a rather similar way.

The substances in soot and oil which cause cancer were isolated by Professor Kennaway of the London Cancer Hospital; and his colleague, Cook, discovered their exact composition. It is now possible to test lubricating oils for their presence, and though cotton spinners still have a high cancer risk, it is probably falling. A specially dangerous occupation is glass-blowing, with a great deal of mouth cancer; and leather dressers and fur workers, for at present unknown reasons, are also particularly susceptible.

A more serious cause of cancer deaths is alcohol. The death-rates in the alcoholic trades (innkeepers, brewers, barmen, etc.) from cancer of the mouth, throat, gullet and so on at ages under sixty-five, are more than double the average. Curiously enough, their stomachs do not develop cancer to an abnormal extent, though they have a high death-rate from gastric ulcers. There can be little doubt that this extra cancer risk is shared by hundreds of thousands of other heavy drinkers. A recent enquiry by Stocks and Karn showed that cancer was particularly common among beer-drinkers, while unboiled milk afforded a good deal of protection. In fact a man is about five times as likely to get cancer if he drinks beer daily and no milk as if he drinks milk daily and no beer.

However, the main cause of excessive cancer is not occupation, but excessive poverty. The Registrar-General divides the people of England and Wales into five social classes, ranging from the well-to-do (about 2½ per cent. of the total) to unskilled workers (about 17 per cent.). As we go down the economic scale the cancer death-rate among men rises even more rapidly than the general death-rate.

When the cancers are classified by their site, we find a sharp distinction. Cancers of the skin, mouth, throat and other organs down to the stomach are about twice as common among the very poor as the rich, with other classes intermediate. Cancer of other organs, such as the intestines, liver and lungs, show no social grading. There are two reasons for this. Firstly, these sites are more exposed to irritation by dirt in the poor than the rich. And, secondly, they are more accessible to the surgeon, and therefore more likely to be cured by operation in the rich than the poor.

Women are, of course, liable to cancer of the womb and

breast. It has long been known that child-bearing makes women more liable to the former and less so to the latter. As rich women have fewer children than poor, they have less cancer of the womb, but more cancer of the breast. However, if we compare the single women, we find that breast cancer has no economic trend, but cancer of the womb increases with increasing poverty.

To sum up, there are a great many causes of cancer which can be dealt with, besides others of which we know nothing at present. Of the causes which can be fought, the most important is poverty (mainly in towns). The second is excessive drinking of alcohol, and in the third place come a group of occupational risks, mainly from soot, oil and pitch.

IRON LUNGS

For the last month[1] the Press has been full of reports of so-called "iron lungs," which are being used to save the lives of victims of poliomyelitis, or sometimes only to prolong them for a few days. Poliomyelitis is a disease of the spinal cord, which runs down from the brain in the hollow of the backbone, giving off a pair of nerves at each joint.

So poliomyelitis causes paralysis by cutting off connections between the brain and the muscles. It is generally called "infantile paralysis," because it usually attacks children. But it can also attack fully grown people—for example, President Roosevelt.

In mild cases, the paralysis is limited to the legs. But if

[1] In the summer of 1938.

it rises higher up the spinal cord the muscles of the arms and chest, and the diaphragm or midriff, a muscle separating the chest and belly, are also paralysed.

Now, when we fill our chests with air we do so by moving the ribs and also by contracting the diaphragm, which forces down the liver and stomach, and makes more room in the chest, so that the lungs expand. Complete paralysis of the breathing is, of course, fatal.

Curiously enough, the heart, which is just as important as the lungs, never gets paralysed in this way. For it goes on contracting even if all the nerves to it are cut. But the muscles used in breathing are also used for other purposes—for example, speaking—and have to be under the control of the brain.

We do not read so much about the epidemic of poliomyelitis in Essex[1] as about the iron lungs, partly because cure is a better news story than prevention, though poliomyelitis is seldom completely cured. Another reason is that poliomyelitis is largely spread by the breath, and is therefore favoured by overcrowding. It is much less likely to spread in a school with small classes than with large ones. So too full an account of the epidemic might give rise to dangerous thoughts (as they are called in Japan) about the effects of economy in education.

The iron lung is the most recent and perfect of a number of methods of artificial respiration developed for use on people who were apparently drowned. The first attempts were made in 1774 by putting a bellows into one nostril and closing the mouth and the other nostril.

This was not very effective. In the nineteenth century

[1] In the summer of 1938

three English physiologists, Marshall Hall, Howard and Sylvester, invented methods which were more effective, and could be performed without apparatus. These were superseded by Schafer's method, invented at Edinburgh in 1903. The apparently drowned person is placed face downwards, and the rescuer kneels astride of his legs and presses on his lower ribs about thirteen times a minute.

This is much less likely to damage him than pressure on the front of the body as in Howard's method. But it cannot be applied for days on end. The first continuous mechanical method was invented by Bragg (now President of the Royal Society) and Paul. This is called the "pulsator," and is a bag which presses rhythmically on the chest and stomach and is very useful in cases of partial paralysis.

The so-called "iron lung" was invented by an American physiologist, Drinker. The patient's body lies in a steel cylinder, in which the air pressure is changed about twelve times a minute. His head sticks out through a rubber collar. When the pressure of the air on his body rises above that outside the cylinder, the air is driven out of his lungs; and when it falls below it the lungs expand again, and air is drawn into them.

Thus the patient can be given food and drink. But from time to time a nurse must go into the cylinder to see to his other needs. The son of an American millionaire has been kept alive in one of these apparatus for several years, and is now said to be beginning to recover control of his breathing muscles.

Of course, the "iron lung" does not replace the lungs, but only the breathing muscles. Since the heart is a pump like the lungs, it is possible in theory to replace it. And this has actually been done in experiments on animals. But in these experiments the great difficulty is that the blood clots in contact with metal, glass, rubber or most other surfaces.

The blood can be prevented from clotting by injecting certain substances. But, if so, it oozes from small cuts. However, I expect that the problem of the artificial heart will be solved in the next fifty years, and a few centuries hence artificial hearts may be as common as artificial teeth.

So far we have been most successful in making substitutes for organs which act mechanically, like teeth and legs, or optically, like the lens of the eye. We are just beginning to replace some which act chemically, like the thyroid gland. This is so because the simpler principles of mechanics and optics were discovered in the seventeenth century and those of chemistry in the nineteenth.

If social conditions allow the continual progress of science, artificial livers, stomachs and other organs should become no more inconvenient than spectacles, and at the same time the progress of preventive medicine should ensure that they are not too often needed.

THE MOTHERS' STRIKE

THE Population (Statistics) Bill now[1] before Parliament will, if it is passed into law, compel everyone who registers a birth to answer a number of questions as to how many brothers and sisters the baby already possesses, how long the parents have been married, how they are employed,

[1] December, 1937.

and so forth. The Government is getting alarmed at the low birth-rate.

Why is this? There are more births than deaths per year, and the population of Britain is increasing. It might be thought that there are enough people in our islands. Nevertheless, there is little doubt that our numbers will begin to diminish within the next five years.

This can be proved as follows. Suppose we consider 100,000 girls born in 1900. Nearly nine-tenths of them lived till 1917. Then a few became mothers. Most of their children were born between 1920 and 1930, and almost all will be born before 1945. If they bear more than 100,000 daughters, the population will increase; if less, it will diminish.

We cannot, of course, calculate the number of daughters exactly for another ten years. But we can calculate, on the basis of the existing death-rates and birth-rates for women of a given age, what this number would be for a group of women who had these particular death-rates and birth-rates. In fact, we can say whether in the long run our population would rise or fall if these rates did not change.

The number of daughters born to 100,000 girls, calculated on this basis, fell from 107,000 in 1921 to 81,000 in 1931. Our population is still growing, because there are a great many women between twenty and thirty years of age. About 1940 it will begin to fall, and may be reduced to half by the year 2000. Even if health conditions were so improved that the death-rate at all ages up to fifty was halved, this would not prevent the decline, though it would slow it down.

The fall in the birth-rate is not peculiar to Britain. It

has occurred in all but one of the states with modern production and hygiene. The exception is the Soviet Union. The Japanese birth-rate has only dropped very slightly. It has fallen in thinly peopled nations like Australia as well as in crowded lands like Belgium.

Until recently reactionary writers deplored the increase of population, and thought that it caused unemployment; though it obviously could not account for unemployment in Canada or Australia. As lately as 1936, Professor MacBride said that the unemployed should be punished by sterilisation for producing unwanted children.

Now their tone has suddenly changed. It is realised that soldiers may be needed for new wars, and that increasing population offers a good excuse for wars of conquest. In fact, as the experience of Abyssinia has shown, conquests do not afford an outlet for "surplus" population. But they do supply labour power at starvation rates for predatory capitalists.

Why is the birth-rate falling? Some people answer that the fall is due to the decay of religion and the use of contraceptives. This cannot be the only cause, because it began falling in Ireland before it fell in any other European country. The fall was at first largely due to emigration of young men. But now that emigration has almost ceased, the birth-rate is still low.

Yet Eire is predominantly Catholic, and birth-control is forbidden. In Bavaria, too, the fertility of Catholic marriages fell more between 1913 and 1920 than that of Protestant or Jewish marriages. If Catholic marriages in Britain are more fertile than the average, this may well be because Catholics are on the whole poorer.

Other writers attribute the fall to the increasing emancipation of women. But in the Soviet Union, where women enjoy a greater equality with men than anywhere else, the birth-rate is not falling. Certainly birth-control and women's freedom are contributory causes. But the primary cause must lie deeper.

Sir Kingsley Wood points out that poverty cannot be the chief cause, because in Britain and many other countries the poor breed quicker than the rich. On the other hand, marriage rates and birth-rates always fall during a period of unemployment. So if we could abolish unemployment, as the Soviets have done, we should almost certainly raise our birth-rate.

I suspect that the root cause is not so much poverty as insecurity. Parents limit their families because they can see no prospects for their children except unemployment and war. Those who have risen a little in the economic struggle are determined that their children should not sink back into abject poverty. They have few children in order that these few should enjoy the education which only money can buy.

In a society which was co-operative rather than competitive, where education was not bought and sold, these motives would no longer apply.

Let us see what other countries are doing to raise their birth-rate. Mussolini has done his best to increase the population of Italy, and has used the increase to justify war. But his best was not very good. Some workers get a bonus of 4 lire (10*d.*) per week per child, but as this is partly financed from a tax on wages it does not go very far. And the birth of the sixth child is rewarded by a portrait of the Duce. In spite of this, the birth-rate is falling steadily, though still fairly high.

Hitler took far more serious steps. A couple gets a loan averaging about £26 on marriage, one-quarter of which is cancelled on the birth of each child. And large numbers of women have been turned out of their jobs. In consequence the marriage rate and birth-rate rose sharply, so that at present more than enough children are born to prevent the population from declining. But the birth-rate is now falling slightly, and the death-rate rising, so the success of Hitler's efforts is not assured.

In France family allowances are made to about 4 million workers, generally as a percentage addition to wages, the average in 1924 being 1*s*. 4*d*. per week per family, though some workers with six children got as much as 18*s*. per week. In no case do the allowances cover more than about half the cost of the child's upbringing, and they have certainly caused no rise in birth-rate.

It is not clear that the British Government enquiry will add anything much to our knowledge, though we shall doubtless get rather more accurate information. A statistical probe is unlikely to reveal the root causes of the fall. And if some system of family allowances is recommended, it will be for the Labour movement to see that it is not used to depress the wages of childless workers and to break the unity of the workers.

A French director of a family allowance fund wrote to an investigator of the Eugenics Society: "From the social point of view, the allowances have prevented the subversive trade unions from using for their revolutionary purpose workers who are fathers of families. The great majority of these have remained outside the class struggle."

The provision of free meals and clothing for school-children would serve the same purpose as family allowances, and serve it much better. But it is unlikely to raise the birth-rate greatly. Until parents can bring children into a world without the risk of unemployment and war, the mothers' strike will go on.

BLOOD TRANSFUSION

In September, 1937, Comrade Jack Kirkpatrick, who had been wounded while fighting for democracy in Spain, lay seriously ill in the Middlesex Hospital. He had lost a great deal of blood and needed a transfusion. Within a few hours forty-two comrades turned up to offer their own. Only two possessed blood of the right kind.

In his diary for 1667 Pepys describes a transfusion of blood carried out experimentally by the Royal Society. The receiver was described as "poor and a debauched man," so from the point of view of the rich and virtuous his life was not valuable. Such experiments were sometimes successful, but often fatal. It was not till this century that Landsteiner in Vienna and Janssky in what was once Czechoslovakia, independently found out the reason for the failures.

Blood consists of a pale yellow fluid, the plasma, which carries round about its own weight of microscopical red corpuscles. These transport oxygen from the lungs to the different organs. The plasma usually clots in a few minutes, but several chemicals will stop this process. Now, if we separate the corpuscles from the plasma by spinning the blood in a centrifuge, as cream is separated from milk, we can put the corpuscles of one man into the plasma of another, and watch the result under a microscope.

Fairly often they become sticky and clump together. In this case the injection of them will lead to grave illness or death. All men belong to one or other of four groups, and the best results are got if both the donor and the recipient belong to the same group. This is not absolutely necessary. I belong to a group whose corpuscles can be injected into anyone with safety, but my plasma will damage the corpuscles of most other people, so only a small amount of my blood can be transferred with safety.

Now these groups are not equally common. In England 46 per cent. of the population belong to group O, 36 per cent. to group A, and only 13 per cent. and 5 per cent. to groups B and AB. Comrade Kirkpatrick was a member of group B. The groups are sometimes numbered, but are generally named after two complicated chemical substances called "iso-agglutinogens A and B," of which neither is found on the corpuscles of group O and both on those of group AB. Corpuscles carrying either of them are damaged when placed in the plasma of a man or woman whose corpuscles do not carry that same substance.

Several remarkable discoveries were soon made. The group to which a person belongs is fixed at birth, and even earlier. It is determined by heredity in a very simple manner. For example, no one ever has an iso-agglutinogen on his or her corpuscles unless one of his or her parents had it. Hence if a man is suspected of being the father of a baby, and the baby's corpuscles carry an agglutinogen not found on his corpuscles or the mother's, he can be cleared of the suspicion. The father must be someone else. If a man got the A substance from his mother and B from his father, he will give A to about half his children and B to the other half, and so on.

Most races include members of all the groups, but the proportions vary. Thus group B is rare in England, but common in India. And it is perfectly safe to transfuse blood from a member of one race into the veins of another, provided they both belong to the same group. It must be very humiliating for a *pukka sahib* in India who prides himself on his English blood to find that there are no Englishmen of the right group available, and that only the blood of a "native" will save his life. Let us hope that Herr Hitler will never find himself bleeding to death with no one but a Jew who can give him the right sort of blood.

The operation of blood transfusion is not difficult. The blood may be drawn straight from a vein into a bottle containing sodium citrate to stop clotting, and then injected still wa m into the veins of the recipient. Or it may be stored on ice for a week or more. The donor should always be tested for freedom from certain diseases. The usual amount transferred at a time is about ½ pint. If more than 1 pint is taken, the donor may feel faint, but I have seen a man cycle home an hour after giving nearly a quart.

I believe that every healthy communist ought to be

ready to give his blood, and not only to comrades. A pint of good red communist blood is better propaganda for the party of Marx and Lenin than a gallon of Tory beer for the National Government. For this reason, it is worth finding out to what group one belongs.

In Spain the blood-transfusion service was well organised. In Madrid a number of donors, many of them women, gave blood once a fortnight. It was stored over ice, and taken to hospitals as needed. A successful transfusion is a wonderful sight. In January, 1937, I accompanied Dr. Bethune, a Canadian who started the first transfusion unit in Madrid, to a field hospital north of the town. A Spanish comrade was brought in with his left arm shattered. He was as pale as a corpse. He could not move or speak. We looked for a vein in his arm, but his veins were empty. Bethune cut through the skin inside his right elbow, found a vein, and placed a hollow needle in it. He did not move. For some twenty minutes I held a reservoir of blood, connected to the needle by a rubber tube, at the right height to give a steady flow. As the new blood entered his vessels his colour gradually returned, and with it consciousness. When we sewed up the hole in his arm he winced. He was still too weak to speak, but as we left him he bent his right arm and gave us the Red Front salute.

BLOOD AND IRON

WHY do we need blood? An average man has about a gallon of blood, and is likely to die if he loses half of it. The blood serves many functions. It carries food materials from the intestine to the rest of the body. It carries water and waste products to the kidneys.

But its most urgent function is to carry gas; oxygen from the lungs to the muscles and other working organs, and carbon dioxide from the muscles to the lungs. Ordinary fluids take up very little oxygen, so that they would be no good for the purpose. Blood contains a special substance, hæmoglobin, which combines with oxygen very easily, and gives it up easily as well. In fact, a pint of blood carries nearly as much oxygen as a pint of air.

Hæmoglobin is of a deep purple colour. If you want to see the colour, prick your finger or ear lobe and let it bleed into some water till you have a nice clear red fluid. Put this in a small bottle and add a crystal of sodium hydrosulphite (not hyposulphite). This will combine with the oxygen and the liquid will turn purple.

Shake it up with air or bubble air through, and it goes red again. This is what happens several times a minute in your own body. The blood in the vein from a working muscle is almost black. Now bubble a little coal-gas through the blood. It goes pink and stays pink, even if you add sodium hydrosulphite. The hæmoglobin is combined with carbon monoxide, and is useless as a carrier of oxygen. If you put your head in the gas-oven your lips will be a nice pink colour after you are dead. But blood that stays red is no use to you, and you will have died because it stayed red.

Hæmoglobin is a protein—that is to say, consists of large molecules much like those found in meat, eggs, cheese and other foodstuffs. But, unlike most proteins, it contains iron. Not a great deal of iron. You have less than $\frac{1}{10}$ oz. of iron in your whole body. Nevertheless, you may easily go short of iron.

You make your hæmoglobin in a very curious place, your bone-marrow. The hæmoglobin is carried about in red corpuscles, which are too small to see with the naked eye, but can easily be seen with a microscope, and just seen with a very powerful magnifying glass.

They last for about a month, and then wear out, and are scrapped in the liver. The iron is mostly carried back to the marrow. The coloured part of the hæmoglobin is thrown out in the bile and finally got rid of in the excreta. If the bile duct is blocked it goes into the blood, and you get jaundice, for the removal of iron and other changes have altered the colour from red to yellow.

If you have too little hæmoglobin you are anæmic. When you work, your muscles cannot get enough oxygen, and you become weak and short of breath.

Anæmia is one of the commonest diseases. In the tropics it is often caused by a small worm, *Ankylostoma*, of which hundreds may find their way into the human intestine, where they suck the blood. This disease is very common in tropical countries where the eggs of these worms live in polluted mud, and bore their way into the legs of bare-footed men, women, and children. This can be completely prevented by proper sanitation and disposal of refuse.

The hook-worm, as it is called in the United States, was apparently brought over by negro slaves. They lived under filthy conditions, and infected one another and their white masters. A few years ago it was found that a large proportion of the whites in the south-eastern states suffer from anæmia due to hook-worms. And it is quite possible that this worm played a big part in winning the Civil War, in which the slave-owners were beaten. The slave-owners kept their slaves without proper sanitation, and the slaves took their unconscious revenge by giving their masters anæmia.

About 1900 there was an outbreak of this worm disease in the Cornish tin-mines, where the climate underground is tropical. It was soon stopped by installing proper sanitation.

In England we are just beginning to discover how common anæmia is. Forty years ago my father invented an apparatus by which the amount of hæmoglobin in a drop of blood could be accurately measured. He bled a number of men, women and children, of whom I was one of the youngest, and worked out the averages. He found that women had less hæmoglobin than men, with very few exceptions.

For a long time doctors thought that this was a natural peculiarity of women, like their smaller average height. But in 1936 Drs. McCance and Widdowson, of King's College Hospital, London, found that women, even in the well-to-do classes, were chronically short of iron. The majority of them made more hæmoglobin if they were given more iron in their diet.

Like many other characters in which women are supposed to be inferior to men, this one turned out to be mainly due to external conditions, and not to be an inborn defect. Women lose some blood every month, and naturally need more iron in their diet than men. But they generally get less.

The iron can be made up in several ways. Many foods contain iron, but, especially in meat, much of the iron is in an indigestible form. The best known sources are liver, cocoa and winkles. But parsley, haricot beans, peas and lentils are also rich in iron. Brown bread and eggs are also good sources. In spite of Pop-eye the Sailor, spinach does not rank very high; and milk, which is otherwise an excellent food, is very poor in iron, though better than beer.

Many medicines sold as cures for anæmia contain digestible iron salts and are therefore valuable. But usually a 2*s*. bottle of medicine contains about a farthing's worth of iron salts, so you will do better to spend the money on liver or cocoa.

Other things than iron are needed to make new blood. A fairly common disease in India, tropical aplastic anæmia, is due to the lack of a substance whose exact nature is at present unknown, which is missing in the diet of many poorly paid Indian workers.

Many other substances besides iron are deficient in a lot of British diets, particularly among the workers. I hope to describe them in future articles of this series.[1] But iron is the very simplest of dietary needs, after water, and a study of our needs of iron is a good introduction to the general theory of diet.

[1] See pp. 194–221.

FOOD

FUEL VALUES

MEN, women and children need food for a variety of reasons. The purposes which food serves may be classed as fuel, growth and repair. Some foods can only serve for one purpose, some for all three.

The simplest requirement is fuel. In order to produce work or heat, a man needs fuel, like a steam engine. Most plants and a few animals can use light as a source of energy. Most animals need a chemical source. None can use electrical energy. And none gets energy from any immaterial origin.

If you put a man in a calorimeter and leave him there for some days on a diet which keeps his weight just steady, the amount of heat that he produces is just the same as the amount which would be got if his food were burned, when allowance is made for the fact that some food is excreted unchanged or partly changed. To be quite accurate, the observed and calculated heat production in a good experiment agree to within one part in 200.

If the man works, 10 to 20 per cent. of the energy in the food may appear as work. Animals give similar results.

Now, at this point some critics will ask me whether I am not saying that man is a machine. Certainly man is a machine in some ways, though not in others. For example, it is one of the properties of a machine that you can replace parts which are broken or worn out. You can replace some parts of a man—for example, the blood corpuscles. But most parts cannot be replaced. For example, you and I may be able to exchange a pint of blood, but we cannot exchange legs. If the experiment is tried on most mammals, the leg lives for a few days or weeks, and then dies. But a leg can be grafted from one frog or insect on to another, at least in early life.

So instead of asking, "Is man a machine?" we should ask, "How much is man a machine?" The answer is that man is less of a machine and more of an individual than a frog or an insect, but still enough of a machine to need fuel.

Now, a steam engine or a petrol motor is a heat engine. All the energy of the fuel is converted into heat, and then some of the heat is converted back into work. Hence an engine can use a great variety of fuels. The steam engine can use coal, wood or heavy oil; the internal combustion engine petrol vapour, alcohol vapour or coal gas.

But an animal is not a heat engine. Some of the energy developed from the union of the food eaten and the oxygen breathed is converted directly into muscular work, as the chemical energy stored in an electrical battery can be converted into mechanical work without passing through the form of heat.

This transformation is a delicate and carefully regulated process. There are dozens of intermediate stages in the oxidation of sugar to carbon dioxide and water. Each intermediate can only be oxidised further if temporarily united with a special kind of protein found in the living cells. These proteins are called "oxidases." Hundreds of different substances which act in this way have now been isolated, and it is clear that the apparently formless slime found in cells which used to be described as "protoplasm" is an organisation of many different sorts of chemical molecules.

How delicately these chemical tools are adjusted to their work is shown by the following fact. The molecules of sugars are asymmetrical, like right-hand gloves or right-foot boots. Chemists can make their mirror images, corresponding to left-hand gloves. But when such "looking-glass sugars" are fed to a rabbit, it cannot use them for food as it can the ordinary sugars. In fact, Alice would have starved to death in the looking-glass world.

The chemical substances which we can use as fuel fall into three groups. The carbohydrates include sugars, and also substances such as starch and inulin (the starchlike substance in artichokes) which change into sugar on digestion. The fats and most animal and vegetable oils are an even better source of energy per ounce, though not per pennyworth. Finally, the proteins, such as make up most of the dry weight of meat, cheese, and egg-white, can be used either for fuel or growth.

A few other substances, such as alcohol, can be used as a source of heat, though probably not of work. The energy or fuel value of foods is easily found, and the number of calories per 1*d.* calculated. Among the cheapest energy sources are sugar (720 calories per 1*d.*), oatmeal (420 calories per 1*d.*), white bread (530 calories) and lard (580 calories).

A man doing light work needs about 2,500 calories

per day. Those doing heavy work may need over twice as much. In Britain (though not in China, India or Spain) most people get sufficient fuel value in their diet, though many do not get it in the most digestible form, nor do they get food which is adequate from other points of view.

The only known animals which need food for fuel purposes only are some adult male insects, such as blue-bottle flies. They do all their growth as grubs, and do not grow new skin and hair, like men. So they merely need food as a source of energy. The females, on the other hand, need body-building foods in order to lay eggs, and become sterile on a diet of sugar and water.

So if men were blue-bottle flies, it would be sufficient to calculate their diet on a basis of calories. But man happens to be less of a machine than is a blue-bottle. And as we shall see, he needs a great many things besides fuel value in his food.

BODY-BUILDING

THE most important function of our diet is to provide fuel. Even in a growing child, only about 5 per cent. of the food is used in growth. Still an animal (except for some insects) cannot live on fuel-foods, such as sugar and fat, alone, though if it gets these only it will live far longer than if it gets nothing. It needs some foodstuffs for body-building and repairs; others in smaller amounts to take part in special chemical processes.

There is no reason in the nature of things why we should

need special body-building foods. Green plants can build up their tissues from carbon dioxide, water, and minerals such as nitrates. Some bacteria and moulds can live and grow on minerals and a fuel-food such as sugar to provide the energy for body-building.

Animals not only rely on plants for their supply of energy, but they trust the plants to do the first steps in building up the chemical compounds of which animal bodies are made. This is still so if they feed on the plants at second hand, or even at fourth hand, like a man who eats a mackerel which has eaten young herrings which have eaten microscopic crustaceans which have eaten still smaller one-celled plants.

The most important solid constituents of our bodies are called "proteins." Muscles, such as ordinary lean meat, consist mainly of proteins, water and salt. Other proteins make up most of such familiar substances as egg-white, cheese and gelatin. These and most other known proteins, though they may form part of living systems, are not themselves alive.

Until last year it looked as if Engels had made a bad mistake when he wrote in *Anti-Dühring* that life was the mode of existence of proteins (or albumens, as the German word used by him is often translated). But since then several viruses have been obtained pure which turn out to be proteins. If you inject one into a suitable plant, the plant becomes ill, and after a few weeks you can extract from it many thousand times as much of this special protein as you injected. So this particular protein does seem to be endowed with a very simple kind of life. But it is exceptional.

Proteins can be split up by acids or digestive juices into about twenty different sorts of chemical molecules which are called "amino-acids." These are among the oddest of chemical substances known. When we pass a current through copper sulphate or quinine sulphate, the copper or quinine moves with the current, because it has a positive charge. The sulphate moves the other way.

Amino-acids are both bases, like quinine, and at the same time acids, like sulphuric or citric acid. So they constitute what is called a "unity of opposites," and manifest new properties. For example, they do not usually travel in an electric field, but they arrange themselves like compass needles in a magnetic field. Proteins which are built up from several hundreds of them have other unique properties.

We can make all the amino-acids in the laboratory, but not in our bodies. Some can be made in our bodies by simplifying others, but at least eight seem to be needed in the food. And some proteins are deficient of one or more of them. Thus zein, a protein found in maize, lacks two amino-acids called "tryptophan" and "lysine." Rats were kept on a diet containing enough minerals, fuel foods and vitamins, but with only zein as protein. They lost weight and died in about a fortnight. If tryptophan was added, they might live for many months, but neither lost nor gained weight. If lysine was added, they at once started growth again.

This dietary was hard luck on the rats, and some people think that such experiments are wicked. Unfortunately they are necessary as long as large human populations are kept on diets which turn out to be unfit for rats.

From such experiments on animals and other less drastic ones on men, followed by chemical analyses of various proteins found in food, we can say whether the protein part of a given diet contains enough of each of the essential amino-acids for the maintenance of an adult, for the growth of a child before or after birth or for the secretion of milk.

The best proteins are of animal origin. But since milk proteins are just as good as meat proteins, there is no physiological objection to a vegetarian diet supplemented by enough milk or cheese. A rigidly vegetarian diet without milk, eggs and so on must be very carefully chosen, and even so may be inadequate for growing children or for nursing or pregnant mothers.

In Spain to-day[1] there is some actual shortage of fuel foods. But even when this is not so, the diet consists largely of bread, beans and other vegetable foods, such as nuts and fruit. Meat is scarce and milk even scarcer. In consequence, many children are dying and adults are showing anæmia. For even if there is enough iron in the diet, the blood cannot be renewed without proper proteins. For this reason, milk is probably the most useful food that we can send to Spain.[1]

Besides fuel foods and proteins, man needs minerals, and special organic substances called vitamins. In Britain at the present time vitamin shortage is the most serious source of malnutrition, and I shall deal with it in further articles.

[1] Spring, 1938. Things became worse later on, and appear still to be very bad.

VITAMIN A, NEEDED FOR SEEING IN THE DARK

UNTIL the beginning of this century many physiologists thought that men could live on a diet consisting of fuel foods to supply energy, proteins for growth and repair, water and minerals. We now know that this is wrong. Besides the amino-acids, there are at least eight other kinds of chemical substance which man needs and cannot make in his own body.

When their need was first realised, Hopkins called them "accessory food factors." But another worker, Funk, who mistakenly thought that he had isolated one of them, called his crystals "vitamin," and the name stuck. Actually the vitamins are a group of very different substances, needed for many different purposes.

Though most of them have now been prepared nearly pure, and their chemical make-up is known, they were originally called by letters. For they were discovered in the following way: Men, rats, guinea-pigs or pigeons were found to get ill or cease growth on a particular monotonous diet. It was then shown that a small supplement of yeast, cod-liver oil, lemon juice or some other food would restore health and growth. Finally, the curative substance was concentrated, but it was given a letter long before it was obtained purc.

The vitamin which happened to be called A is a red, greasy, crystalline stuff which melts to an oil when warmed. It is found in the liver of many animals, particularly fish. Though animals cannot make vitamin A from simple constituents, they can easily make it from carotene, the orange-coloured oil which gives its colour to carrots, and is found in green leaves, yellow maize and many other plants. The vitamin A formed from carotene by oxidation is stored in the liver, and a certain amount is secreted in milk. But naturally a cow which is fed on oilcake does not produce so much as if she gets green grass.

Who discovered vitamin A? In 1888 Lunin in Switzerland first showed that some unknown chemical was needed in the diet. In 1912 Hopkins in England came to the same conclusion on far more critical evidence. In 1915 McCollum and Davis in America showed that at least two vitamins were needed. In 1928 Euler in Sweden found that carotene could replace it in the diet. In 1930 Moore in England showed that animals make the vitamin from carotene, and finally[1] the story goes back to Switzerland, where Karrer determined its chemical formula in 1931.

Hundreds of other workers helped in the discovery, which was a social, not an individual act. Its difficulty arose from the tiny quantities of vitamin which are needed. A single ounce of it would restore growth to normal in 30 million rats, or perhaps 1 million babies.

Why do we need vitamin A? The most serious symptoms due to its lack are a drying up of the skin and still more of the delicate membranes covering the eye and lining the mouth, throat, intestines and other organs. These are then attacked by a variety of microbes which are harmless to properly nourished people. About half the

[1] Meanwhile two Soviet biologists have found a closely related substance, vitamin A_2, which seems to be equally useful, in fresh-water fish.

blindness of children in India is attributed to vitamin A shortage. The amount of slight skin infections, such as "napkin rash," in a group of London babies was nearly halved when they were given extra vitamin A.

A prolonged shortage affects the nervous system and slows down growth. But the symptoms are not always obvious, and until recently many scientists thought that most British children got enough of it.

Then a dramatic discovery was made. The retina, as the sensitive film at the back of the eye is called, adapts itself to darkness when we go into a dark room from daylight. The process takes about an hour to complete, and is due to the formation of a light-sensitive substance called "visual purple." This has been found by Wald in the U.S.A. to be a compound of vitamin A with a protein. And people who are short of vitamin A, even if their skin is normal, cannot see well in the dark.

Children are tested as follows. They look at a bright white screen for some minutes. Then the light is turned out and they are asked how many of a series of dimly lit patches of light, some brighter than others, they can see. The test is repeated after ten minutes in the dark.

Harris and his colleagues found that 90 per cent. of well-fed boys in a so-called "public" school reached the standard regarded as normal. But the proportion of children in three elementary schools in London and Cambridge with proper dark vision varied between 40 and 45 per cent. Similar differences were found between the mothers of the elementary school-children and well-fed adults.

Fifteen drops a day of halibut-liver oil improved the

dark vision of thirty-nine out of forty children, and brought it back to normal in thirty of them.[1] Things would almost certainly be worse in the distressed areas, and probably half our population is short of vitamin A.

How many coal-miners and lorry drivers, to mention two groups who need the finest possible vision in darkness, are short of this vitamin? It seems likely that a good many accidents, both underground and on the roads, could be prevented if the workers concerned were properly fed. If the question of diet were brought up at the inquest on every victim of an accident occurring in the dark, it might be possible to arouse the public conscience in this matter.

A diet with plenty of green vegetables is sufficient as a source of this particular vitamin. Jeans and Zentmire, who invented the visual test for vitamin A, found that in Iowa most of the children of quite poor agricultural workers had enough of it, while town children generally lacked it. But the artificial diet of town workers is easily supplemented by fish-liver oils—if they can afford them.

At the same time that Maitra and Harris were curing English school-children, a German scientist announced that partial night-blindness was hereditary. This is probably not mere Nazi propaganda. It is likely enough that some people need more vitamin A than others for proper night vision, and that the capacity for seeing with a minimum of it is hereditary. Perhaps 1,000 years of eugenical sterilisation might produce a race which could

[1] Later evidence suggests that vitamin C is also needed, and perhaps the two together would have cured all the children.

see in the dark even on a German worker's diet!

But biologists like myself, who would actually like to see some improvement in conditions in our own time, prefer to change environment rather than heredity, and demand that where scientific standards of diet are clearly known, they should be enforced.

VITAMIN B

THE diet of every nation has its strong and its weak points. In England the poor were once so short of vitamin D that doctors in Europe called rickets the "English disease." Two other deficiency diseases are rare in England, but one, beri-beri, is common among rice-eaters. The other, pellagra, is widespread where maize is the chief cereal eaten.

Beri-beri is the Malay word for a disease common in India, Java, China and Japan. Its first symptoms are numbness, pain and later paralysis of the legs. The paralysis may spread and the heart is affected. The limbs may or may not swell, and death is a very common outcome.

On a Japanese warship cruising round the world about 1880 half the crew fell ill, and one in fifteen died. A naval doctor called Takaki put the disease down to bad food, and found that it disappeared when the rice diet of the sailors was supplemented with meat, fish and other extras. Beri-beri disappeared in the Navy, and Takaki was made a baron. Unfortunately, beri-beri is still found among

Japanese workers on land. And those who say that they are underfed are not made barons, but imprisoned for "dangerous thoughts."

The next step was made by a Dutch doctor, Eijkman, in Java, where the disease was very common. He found that it occurred in hens as well as men if they were fed on milled rice. But the hens recovered if they were given the bran of the rice; and so do men.

Later McCollum and Davis found that rats need something found in rice bran, and other foods, if they are to grow, and to escape paralysis. This was called "vitamin B." For over ten years it has been known that this account was much too simple. The old vitamin B extracts include the substance curing beri-beri, a quite different one curing pellagra, and at least two others.

No one would dream of advertising in a scientific journal that a particular food is rich in vitamin B. But an advertisement of that sort is good enough for the ordinary consumer, and will be until some systematic effort is made to spread scientific knowledge among the people.

The preventive of beri-beri was isolated by two Dutch chemists named Jansen and Donath. It is an organic base containing sulphur, and now called "aneurin," "thiamine," or sometimes "vitamin B1." It plays a part in speeding up one of the chemical processes that go on in nerves and muscles.

Pellagra is a disease whose most obvious symptom is a skin inflammation. But the intestines are often inflamed, too, and the victims fairly often go mad. In 1927 there were 120,000 cases in the southern United States, of whom 5,000 died. The deaths rose to 7,000 per year during the depression. Almost all of them were Negroes. It is also common in Egypt, and not rare in Italy and Rumania, while at least up till 1932 it was found in Georgia and Tashkent.

Goldberger in America showed that it was not infectious by trying to give it to himself by inoculating blood and other fluids from victims under his own skin. He showed that it could be cured by proper diet, especially if this contained fresh meat, milk and eggs. But he could not isolate the vitamin.

It was finally discovered in a very strange way. Warburg in Berlin found that certain bacteria contain a yellow ferment with which they oxidise some of the substances in their food. Later this was found in many animals and plants. It can be purified, and will then cause the oxidation (or, rather, one particular step in the oxidation) of many times its weight of sugar per minute.

To do so, however, it needs still another chemical found in yeast juice. The ferment and the co-ferment were both purified. The first consists of a protein to which is attached a yellow substance called "lactoflavin," found in milk and many tissues. Kuhn at Heidelberg showed that this was a vitamin, one of the four or more different substances in the old so-called "vitamin B," all of which are needed for growth.

The co-ferment is a compound, including, among other things, nicotinic acid, an acid originally made from tobacco. A biochemist named Knight at the Middlesex Hospital, London, was trying to grow the bacteria causing boils, on simplified food. He found that they, too, needed small quantities of something, and that the something was nicotinic acid plus aneurin.

Finally, within the last three months Elvejhem in Wisconsin and Harris and Hassan in Egypt showed that nicotinic acid will cure pellagra. And Euler in Stockholm made it probable that the actual vitamin is the same as the co-ferment discovered by Warburg. So Warburg had discovered two vitamins without knowing it.

There is still a fourth substance, and perhaps more, to be discovered in what used to be called "vitamin B," and it will probably be some years before we learn how much of each of them a growing child needs.

There is no clear evidence that disease due to lack of any of them is common in England. Nevertheless, some cases of neuritis have recently been shown to be suffering from a shortage of aneurin, and part of the benefits got from extra milk in children's diets may be due to the lactoflavin in the milk. A diet largely consisting of white bread, margarine and potatoes certainly contains too little of several of the B vitamins.

Now that they have been obtained pure, it should be possible to get much more definite information. And here the Central Institute of Nutrition in Moscow, with 125 beds for human patients, should furnish data as accurate as those obtained at Cambridge by work on rats under the Medical Research Council.

There is no question but that millions of people in India are terribly short of aneurin. Beri-beri is common in children and pregnant mothers in the rice-eating districts of Southern India, and is likely to continue so until real wages are raised, though this condition is partly due to the introduction of factory methods of rice-milling, which remove all the bran. Since the British introduced these methods, along with other features of capitalism, they are responsible for this disease, and should either take steps to remedy it or make way for others who will do so.

VITAMIN C AND SCURVY

When we read of the great voyages of discovery by which, from 1400 to 1800, Europeans reached almost all the world's coasts, and opened the way for imperialism, we think of the storms and hidden rocks which the explorers braved. But at least as deadly an enemy was a disease called "scurvy."

When Vasco da Gama, the Portuguese navigator, sailed round the Cape of Good Hope, 100 men out of his crew of 160 died of scurvy. Doubtless some Indians wish that the other sixty had died too, for it was by this route that Europeans first reached India.

What was this scurvy? The symptoms are great weakness, swelling of the limbs, and, above all, a brittleness of the small blood vessels, which causes them to burst, so that the whole body may be covered with red or purple spots. The joints become very painful, and the teeth may fall out. The disease is fatal if not properly treated.

The cure was known more than 300 years ago. Sir Richard Hawkins, the founder of the slave trade between

West Africa and America, knew that oranges or lemons were effective. And in the eighteenth century Captain Lind, of the British Navy, proved this conclusively by experiments on sailors.

Thirty years ago it was generally admitted that scurvy was due either to something lacking in the diet or to tainted food. But there was one place where other views were held. Osler wrote in his *Textbook of Medicine*: "In parts of Russia scurvy is endemic, at certain seasons reaching epidemic proportions; and the leading authorities in that country are almost unanimous in regarding it as infectious." I do not suppose that a doctor who said that the Russian peasants suffered from malnutrition would have risen to be "a leading authority" under Tsarism.

Most animals do not get scurvy, however badly they are fed. They can make their own vitamin C, as the anti-scurvy factor is called. But monkeys and guinea-pigs need it like men. And when Holst and Frolich produced scurvy in guinea-pigs in Norway in 1907 it soon appeared that it was due to lack of a substance needed in small amounts; and the quantities in different foods could be roughly measured in doses needed to cure a guinea-pig.

I was in the next room when Szent-György, a Hungarian working at Cambridge, isolated vitamin C. He was not looking for it, and did not know that his crystals were the vitamin. He had tracked down and purified a substance found in various plants and concerned in oxidations inside the cells of both plants and animals, where it plays a part somewhat like that of hæmoglobin in the animal body as a whole.

At the same time two German chemists called Tillmans and Hirsch, of Frankfurt, were on the track of faked fruit juice substitutes. They found that natural fresh fruit juices bleached certain dyes, but stale ones and artificial substitutes did not. Later on they suggested that their bleaching substance, Szent-György's crystals, and vitamin C were the same thing. Within a few months Szent-György, now back in Hungary, and King in America showed that this was true.

Finally, Hirst in Birmingham worked out the formula of the vitamin, now called "ascorbic acid," or "cevitamic acid," and it was made from milk-sugar by a complicated process, simultaneously in Birmingham and Zurich. I have left out important work done in France, and several other countries, which contributed to the final result, which showed clearly enough that discovery is a social process and an international one. Szent-György thinks that still another substance is needed to prevent scurvy. But if so this is present in sufficient amounts in most diets.

Scurvy used to be very common in England every winter, but died out as a serious disease during the eighteenth century, though there were a few cases during the last war. This is because, although ascorbic acid occurs in many foods, it is readily destroyed by heat and air. So workers who lived on bread and salt meat throughout the winter succumbed to it.

Potatoes contain a lot more than bread, even after cooking. And their introduction was one factor in wiping out serious scurvy. The potato shortage in 1917 caused outbreaks of scurvy in Glasgow, Manchester and Newcastle. As Engels said, the potato was the biggest addition to the raw materials at man's disposal since the discovery of iron. Another blow to scurvy was the introduction of root crops to feed sheep during the winter, so that it is not necessary to slaughter them in the autumn and salt their carcasses. But mild scurvy is still regrettably common in babies, showing itself by painful and often swollen joints. A little orange juice added to the milk will prevent this. How much "rheumatism," bleeding gums, bad teeth and other illnesses in adults are due to mild scurvy we do not know. Chemical tests show that many British workers are near the borderline of vitamin C deficiency.

The best sources of ascorbic acid are oranges, lemons and grapefruits. Some apples are rich in it. So are tomatoes; but bananas and grapes are not so good. Green vegetables, such as watercress, are as good as oranges, but much of the vitamin in cabbages is destroyed by cooking, especially with soda. Most preserved and canned foods contain none at all.

If scurvy is not a serious problem in Britain, it is so in many parts of the British Empire. In South Africa it is actually beginning to attract the notice of the whites, because, to quote a recent report, it "is one of the important factors limiting the employment of the native in industries." It is also a problem in the Canadian Arctic and in several other regions.

Still it is pleasant to be able to record a substance of which most people in Britain, even among the worst paid, are not grossly deficient, and of which they certainly get more than did their ancestors.

VITAMIN D, THE PREVENTIVE OF RICKETS

Most of the people of England would have better bones and teeth if they had been better fed. A small proportion of them have skeletons so deformed that they are said to suffer from rickets. In its extreme form it shows itself in bow legs, hunched back, swollen joints, and other deformities, both in growing children and pregnant women.

These were common enough fifty years ago. They are rarer now, but as lately as 1928 the Ministry of Health reported that 87 per cent. of 1,635 unselected five-year-old London school-children showed some sign of rickets, often more or less healed, but never quite brought back to normal.

In most cases the results will not be terrible. A few more decayed teeth, another half-hour of labour pains, a bone that broke under a strain which a thoroughly healthy bone would have stood. But some of the girls will die in childbirth because their pelvis is too narrow, or bear babies doomed to paralysis from head injuries. Some of those decaying teeth will start a fatal infection. Few of those children will grow up as happy and as useful as they might have done.

What is the cause of rickets? It is common, and was commoner, in Northern Europe. In Southern Europe and the tropics it is found in cities, but rarely in the country. So many doctors ascribed it to lack of sunlight. On the other hand, cod-liver oil was a popular cure for rickets, and many doctors had found it a useful remedy.

So twenty years ago there was a fierce controversy as to whether rickets were caused by lack of sunlight or bad diet. It seemed impossible that both explanations could be true. But they were. In 1919 Huldschinsky cured a number of German babies who had got rickets as a result of the British blockade with ultra-violet radiation from special lamps. As he was a Jew, he would not be allowed to cure German babies to-day of the effects of Goering's substitution of guns for butter. But at the same time Mellanby in Sheffield showed that rickets could be produced in dogs given a monotonous diet, and cured by extracts of cod-liver oil.

Darwin, who was a great experimenter, occasionally tried what he called a "fool's experiment." He was so impressed by the sensitivity of climbing plants that he once made his son play a bassoon to a convolvulus to see if it would be affected, which it was not! In 1924 two Americans, Steenbock and Hess, both tried a "fool's experiment," the sort of experiment which one does not publish unless it succeeds.

They gave rickets to some rats, and then shone ultra-violet rays, not on the rats, but on their food. The rats recovered. They then separated the various food constituents, and played their rays on each in turn. It was not till 1926 that Rosenheim and Webster in London, and Windaus in Munich, simultaneously found that the substance which was turned into vitamin D was a wax called "ergosterol." It has since been shown that several other closely related chemicals can also be made into substances which cure rickets.

The contradiction was now solved. A child can get its

bottled sunshine in eggs, butter, fats or fish-liver oils. Or it can make vitamin D by exposing its own skin to sunlight or the right kind of artificial light. No doubt sunlight is the cheapest preventive, but it is doubtful whether an English child, even if it runs about through the summer in bathing drawers, can store enough vitamin D to last it through the winter.

And a Negro or Indian child in England certainly cannot. The coloured peoples are protected against blistering by a tropical sun, but are very liable to rickets when the sunlight is milder. In fact, in England the whites are "superior race" to the Negroes, who are liable to rickets. And in West Africa the Negroes are the "superior race."

Ergosterol and vitamin D belong to a group of substances called the "sterols," which have a variety of strange effects on living creatures. They include some hormones which are needed for life, and others which first appear in the body at puberty and cause it to change. Ergosterol was first discovered in the drug ergot by a French chemist, Tanret. It was found to have no action as a drug, and he probably thought he had wasted his time in isolating it. He certainly never dreamed that it was concerned in preventing rickets.

How does it work? No one really knows. But we do know this. The immediate cause of rickets is a shortage of phosphate or calcium, or both, in the blood. These are needed for bone-building. And in a healthy child the blood contains nearly twice as much phosphate as in an adult. But if the adult breaks a bone his blood phosphate goes up, provided he is not too old and has plenty of vitamin D.

We suspect that vitamin D somehow helps the intestine to absorb calcium phosphate from the food. But we don't know a bit how this is done, as, for example, we know how vitamin A is concerned in night vision. If you give an animal or a baby too much of this vitamin, the calcium and phosphate in the blood rise to too high a level, and stones are formed in the kidney, the arteries harden, and so on. However the effects soon pass off when the dose is stopped. Nevertheless, an overdose should be avoided, whereas none of the other vitamins seem to be harmful, however much you take.[1]

There is no danger from cod-liver oil, which contains about one part in a million. But some of the preparations made artificially by shining ultra-violet radiation on ergosterol are so powerful that the stated dose should never be exceeded.

We have still a lot to learn about vitamin D, which lays traps for the unwary scientist. For example, I predicted that as a result of sunshine there should be an annual tide in the blood phosphate of adult men and women, the highest level being reached in July or August, the lowest in February or March. My pupils, Havard and Ray, analysed their blood and that of others throughout a year, and found the cycle which I had predicted.

When February came we were at the bottom of the slump, and I said, "Now we will just prove the theory by bringing our phosphate up to summer level." We sat under ultra-violet lamps till we were brown. We ate vitamin D in amounts which would have cured a rickety

[1] Since this was written, the effects of an overdose of nicotinic acid, the preventive of pellagra, has been shown to cause illness.

child, and nothing whatever happened! I had made a true prediction for a false reason, and the annual cycle is and remains a mystery.

But there is no mystery about the effects of the economic cycle. Between 1929 and 1933 the percentage incidence of malnutrition rose six times. In 1934 the frequency of rickets in London school entrants, which had been falling for some years, rose again. It is no use discovering how to make vitamin D if those who need it cannot afford it.

MINERALS IN FOOD

BESIDES organic constituents in food—that is to say, complicated carbon compounds—we need inorganic substances—that is to say, water and minerals. Some are needed in very large amounts, some in the tiniest traces.

Our greatest need, after water, is sodium chloride, or common salt. We need it for a very curious reason. It is not found in most cells of our body, nor in most plant cells. They contain potash salts, and there are plenty of these in every kind of food. But it is found in our blood.

The simpler sea animals have no blood at all. Sea-water penetrates all through a jelly fish or sea anemone, and their cells are accustomed to it, and die if its composition is much altered. Most sea worms, and also molluscs such as the oyster and whelk, and crustaceans such as the lobster, have a blood very like sea-water, but also containing foodstuffs which are carried round to the cells.

So do the most primitive fish, the hagfishes. In the

majority of fish, and in all land animals, the blood corresponds to sea-water diluted with about three times its volume of fresh-water. And the lives of some babies which had lost much water and salt from diarrhœa have been saved by injections of diluted sea-water. However, an artificial mixture is safer.

It is an amazing fact that the heart of a dead rabbit, or for that matter a dead man, will generally start beating again if it is taken out within an hour or so of death, and warm water containing oxygen and the right mineral salts is run through it.

Besides common salt, a little of a lime salt is needed. London tap water contains about the right amount. Without this the heart will not contract. And without a trace of potash salts it will not relax. Though we live on dry land our cells still live in a kind of brackish sea-water to which some foodstuffs have been added. And it is worth remembering that we pass the first nine months of our lives in a watery environment.

A dog, a rabbit, or a bird only needs salt for this purpose, and can go for a long time without it. But men, horses and cows, among others, need salt for another purpose. We cool ourselves by sweating, and sweat contains a good deal of salt. But many warm-blooded animals do not sweat. The dog and cat have no sweat glands except on the bare skin of their feet. The need for salt is particularly felt by vegetarian animals which sweat, for there is very little salt in most plants.

Our horses and cows are often short of salt. That is why they lick one another in summer. The only horses which I have ever seen which got all the salt they could possibly want were the pit ponies in a Cheshire salt mine. They had licked great holes in the wall of their underground stable.

Men who sweat a great deal feel an instinctive need of salt. Miners in deep and hot mines eat far more bacon and kippers than the average of the population, and some of them put a little salt in the drinking water which they take underground. If they run very short of salt they may get cramp in the limbs or stomach.

The same applies to other workers who sweat very greatly, such as ship's firemen. The firemen of Scandinavian ships eat more salt fish and salt meat than those of British ships, and are therefore less affected.

But the need of salt is most felt by vegetarians in hot countries such as India. Here it is a necessity of life. In England it is somewhat of a luxury, and most of us eat more of it than we need, though it probably does us no harm. It could be taxed without injustice. But in India the salt tax weighs most heavily on the poorest workers, and Mr. Gandhi's campaign for its abolition was biochemically justified.

Besides common salt, we need lime salts and phosphates to build bone, particularly during childhood and during pregnancy and nursing. Milk, though poor in iron, is a good source of both these elements. Though rickets are probably more often due to lack of vitamin D than of lime, the latter is undoubtedly short in the diets of the poorer British workers.

In many districts goitre, a swelling of the thyroid gland in the neck, is common. Indeed it used to be known as "Derbyshire neck," and it was so common in the Great

Lakes basin of North America that in 1917 the mobilisation of conscripts in Michigan was delayed because there were not enough uniforms with outsize collars. Goitre is also common in Switzerland and some of the Himalayan valleys.

Now the thyroid gland produces a special protein containing iodine, which is needed by the rest of the body. And if one is short of iodine its swelling is one symptom of general ill-health. Very small quantities of an iodide added to common salt prevent the appearance of endemic goitre, and often cure it. It was only possible to raise sheep in some parts of Michigan when they were given salt containing iodides to lick.

It is, however, worth pointing out that most cases of goitre in England are not due to iodine deficiency, and an attempt to treat them with iodides is likely to do more harm than good.

Iron deficiency is a fairly common cause of anæmia, but small amounts of copper, zinc, manganese and cobalt are also needed. These are usually present in ample quantities. But the soil in parts of California needs more zinc if orange trees are to grow there, and there is an area in New Zealand where sheep cannot live because of the lack of cobalt.

Every farmer knows the importance of soil for plants. But only now is its importance for animals and men being discovered, and nowhere is better work on this topic being done than under Sir John Orr at the Rowett Institute near Aberdeen.

In 1866 Marx was greatly interested in Trémaux' theory that soil determined racial differences. Engels convinced him that Trémaux had gone too far. But though Engels was certainly right, Marx had a better case than appeared at the time, and we may still discover that some of the supposed inborn differences between human races are really due to geological differences in their homes.

DRUGS

GERM-KILLERS

ONE thing which I particularly liked about the Soviet Union when I visited it was the almost complete absence of propaganda as compared with England. I don't mean political propaganda, but biological propaganda, propaganda which disgusts me as a biologist. In the advertisement columns of the newspapers and on the hoardings we are told the most fantastic biological tales. For example, that it is dangerous to have acid in your stomach, that pains in the back are generally due to kidney disease, or that your skin is healthier if you smell like a tar-barrel instead of a human being.

In truth, you need strong acid in your stomach for normal digestion; and pains in the back are generally due to inflammation of the back muscles, displacement of the joints in the back, or disease of some other internal organ than the kidney. And while an occasional bath is very desirable in order not to get lousy, too much washing with soap appears to remove a protective film of grease which is secreted by special glands in the skin.

This propaganda is even more obviously put over for economic reasons than the very similar propaganda about "our" Empire, and "our" flag, which by the way, no longer serves to protect British seamen, as it once did. In this country one can still reply to some of the political propaganda. But one cannot reply directly to the biological propaganda. I was brought up on the comic theory that our Press is free.

About fifteen years ago I was writing a book in which I mentioned the analysis of Beëcham's Pills which had been made on behalf of the British Medical Association, and published in their book, *Secret Remedies*. The publishers at once cut the passage out. They did not want to risk an action for libel, though I never suggested that these pills contained any harmful substance, but only gave the estimated price of their ingredients. So in this article I shall only be able to write a very tiny fraction of the truth.

The medicines which you buy at the chemist's are of two kinds. They may be of secret composition, in which case you will generally find a stamp for taxation purposes on the packet. This sort of medicine usually contains a well-known drug puffed up as a wonderful secret discovery, but sometimes consists of sugar and water with a pink dye, or diluted vinegar. A few are definitely harmful. Some of the slimming cures act by producing such violent indigestion as to force you to starve yourself.

The other medicines give their composition, and many of them contain well-tried remedies, which work in a great many cases. The formulæ are often less mysterious than they look. Thus "Koray," which is well known to readers of the *Daily Worker*, consists mainly of acetyl-salicylic acid, put up with starch (amylum), cane sugar (sucrose) and a small amount (q.s. is an abbreviation for the Latin *quantum sufficit*, or "as much as is needed") of tetrabromfluorescein, a red colouring matter. You actually get about half as much again of the acetyl-salicylic acid, which is the active constituent, per tablet as in some other preparations of the same kind.

The mystery is partly the fault of the medical profession. They have a series of most impressive-looking symbols for weights and measures, and use Latin words for the most ordinary substances. For example, glycirrhiza means liquorice, and if you ask your doctor to prescribe for a child's constipation, he is likely to prescribe "Extractum glycirrhizæ liquidum" which is much more impressive than "liquorice and water," and also costs much more.

In spite of this, about once in five times the doctor finds something seriously wrong, with which he can deal, and with which you could not deal yourself. So we cannot dispense with doctors just because they do not always tell us the whole truth.

In fact, there is a medical racket and a "patent medicine" racket. Each fiercely denounces the other. The doctor tells you to avoid proprietary articles, and often prescribes the same substances himself. The patent medicine merchant tells you to avoid harmful drugs and trust to natural remedies, although he may be selling a natural product which is quite a powerful drug and harmful if you take too much.

This racketeering is inevitable under capitalism for two reasons. In the first place there are big profits to be made out of selling either genuine or bogus medicine. Secondly, the general public is not taught medicine or pharmacology. This need not be so. Lenin looked forward "to the education, training, and preparation of people who will have an all-round development, an all-round training, people who will be able to do everything." And it is very much more important that everyone should be something of a doctor and a pharmacist, and able to do "running repairs" on himself or herself than that they should be able to do so to a motor car or even a bicycle.

Unfortunately our first-aid classes do not go far enough. We should not think much of a motorist who had never looked under the bonnet of his car. But we should mostly be horrified at the idea that every child should look at the inside of a man and of a woman. Yet till this is done we shall find it difficult to take a sensible and objective view about our health.

What ought everyone to know about drugs? First of all, they can be divided into three great classes: those which kill parasites, visible or microscopic, those which destroy poisons and those which act on our bodies. Of course, some drugs have a double-barrelled action, and others, such as hypophosphites, have none at all, except the economic effect of transferring money from the patient or his approved society to the chemist and manufacturer.

The danger with germ-killing drugs is that they may kill the patient as well as the germ. For example, antiseptics such as phenol ("carbolic acid") and mercuric chloride ("corrosive sublimate") are very valuable for killing germs on the skin before an operation. And they are used in treating infected wounds. But they may be absorbed from wounds or when swallowed, and have killed a number of people.

Among the most important antiseptics for external use are those like calomel ointment and protargol, which can be used to prevent the transmission of venereal diseases from one person to another. They are outrageously expensive, and it is illegal to sell them with instructions for their use, so these diseases are commoner than they would be if the antiseptics were correctly employed.

Other drugs act on parasites inside our bodies. Worms are no great danger in England, but kill millions of people in the tropics. All the drugs which kill them, such as antimony tartrate and male fern extract, are poisonous to man in large doses, and they should never be used without medical supervision. Quinine kills off the parasite causing malaria, emetine (from ipecacuanha) that of amœbic dysentery, and salvarsan that of syphilis. These are all of them single-celled organisms which can wriggle about, and are probably best considered as animals, while most infectious diseases are caused by bacteria, which are regarded as plants, though the distinction is not very sharp.

Until recently there were very few ways of killing bacteria in the body unless they were concentrated in a small wound or abscess or in some special region such as the urinary tract. But recently an immensely powerful group of drugs, including sulphanilamide and its derivatives, have been put on the market. The best known of these has the trade name Prontosil. This has cured a number of cases of septicæmia ("blood poisoning" and puerperal fever) in the most dramatic way. It also acts on local infections, such as gonorrhœa and some kinds of abscesses. But it is a dangerous drug, and has killed a number of people.

The ideal germ-killer has not been found, and no germ-killing drugs, except quinine, should ever be taken by the mouth except under supervision, still less injected. For the dose needed to kill a man is only a few times greater than that needed to disinfect him.

PAIN-KILLERS

In the last article I wrote of the very few drugs which actually kill germs in the body. Still fewer neutralise the poisons produced by them. This can be done by anti-toxins in a few diseases, such as diphtheria; but they have to be injected, and are quite useless when taken by the mouth.

Almost all drugs are taken for their action on the human body. And the vast majority of those which are of any real value will kill you if you take five or ten times the dose which is needed to produce the desired effect. This is so, for example, with morphine and digitalis, which are therefore scheduled as poisons and only administered by a doctor or nurse.

Probably the medicines which are most widely taken are purgatives, whose use has become a habit with many people. One very common type are the so-called "liver salts," "fruit salts" and the like, which are sold with various exaggerated claims at enormous profits. They all act in the same way. They cannot be absorbed from the intestine into the blood, and are therefore got rid of. They do not "purify the blood," as often claimed, because they are not absorbed into it.

In nine cases out of ten, the reason why these things are needed is a social reason. Many of us do not take enough regular exercise, and also eat a diet which leaves too little residue behind. So a return to a more natural life, in which every healthy adult did even an hour's hard work daily, and our diets contained plenty of vegetables and fruit, would ruin the "healthsalt" merchants. Meanwhile, one can buy enough sodium phosphate or potassium bitartrate at a chemist's to do what is required for a very small fraction of what one pays for proprietary articles.

Besides these, a great variety of substances are taken for the same purpose which act by irritating the bowels. Phenolphthalein, which is sold in about half a dozen "patent medicines," is usually fairly harmless, but occasionally causes a strange skin eruption, with purple or pink patches which will last for years. Needless to say, this fact is not advertised.

Another group of drugs is sold for "indigestion"—that is to say, burning or gnawing pains after meals, especially meals rich in sugar. These drugs mostly consist of magnesia, which neutralises the acid juice in the stomach which is normally there, and only irritates you if your stomach is in some way abnormal. Thus, although they relieve pain, they do not take away its cause.

The commonest cause of gastritis—that is to say, an inflamed and irritable stomach—is worry and anxiety. It is particularly common among busmen and travelling salesmen. I had it for about fifteen years until I read Lenin and other writers, who showed me what was wrong with our society and how to cure it. Since then I have needed no magnesia. But these pains may also be due to gastric ulcer or even cancer. So it is better to consult a doctor, even though he will probably recommend magnesia. But the *Daily Worker* may effect a permanent cure.

The large majority of drugs act on the nervous system. A great many of them depress its activity. Some of them, such as chloral, sulphonal and veronal, lead to a condition not unlike natural sleep, and are given for insomnia. But in the long run they may all lead to habit formation.

The great chemical trusts which produce these drugs use a propaganda upon the medical profession which is often just as unscrupulous as that of the "patent medicine" sellers against the general public. New drugs are constantly being turned out which differ very little from the well-known ones, but are advertised as much safer, until they in turn are found out.

Two of these sedatives can be bought at any chemist's, and are worth having in the house. One is sodium bromide, which was till recently the standard treatment for epilepsy, and is a valuable remedy for "nerves," in the sense of jumpiness. It is also a fairly good prophylactic for sea-sickness. It will often relieve sleeplessness. And it has a great advantage. If you take too much you feel miserable and depressed and may come out in very nasty spots. So it very rarely forms a habit.

Paraldehyde will give sleep as effectively as chloral or veronal. And, unlike chloral, people with heart disease can take it. But its taste and smell are so unpleasant that no one is likely to take it for fun, and it is not often a drug of addiction.

Most drugs which act on the nervous system paralyse some particular part of it. For example, when you bend your elbow you are doing two things at the same time—namely, contracting the biceps muscle of your upper arm and relaxing the triceps muscle on the other side of the bone, which pulls the joint in the opposite direction. Now, if you are poisoned with strychnine the relaxation does not occur. On the contrary, the two antagonistic muscles pull in opposite directions and you get an appalling cramp.

In just the same way the upper parts of your brain are largely engaged in inhibiting the activity of the lower parts, which are concerned with simple activities and simple emotions, such as rage. Many drugs act in the same way as alcohol, and paralyse the higher parts of the brain to a greater or less extent, so that you tend to behave less like a man and more like an animal. This may be quite a good thing if it does not go too far, as it is quite good to take an occasional holiday and live like a healthy animal for a few days.

Many of these depressants also relieve pain. Morphine and heroine are particularly useful in this way, and the latter is remarkably useful in dealing with the intractable tickle of a severe cough. Unfortunately, some people like their psychological effects. I do not. I have taken a large dose of heroine four times a day for ten days or so without getting any "kick" out of it or losing an hour's sleep when I stopped taking it.

And I suspect that in a decent society, where no one

wanted to take refuge from reality in dreams, these drugs could be sold openly as pain-killers. The Japanese have used morphine and heroine with considerable effect to demoralise the Chinese population. I very much doubt whether they would be able to make many addicts among people under forty years old in the Soviet Union, even if they were allowed to try.

So while it is quite right for the League of Nations to fight against the trade in such drugs as these, two things are worth pointing out. They are not habit-formers with everyone. And their illicit manufacture and sale brings in such huge profits that it is most unlikely to be suppressed in those parts of the world where the whole social system is based on profits.

In countries where there is an illicit drug trade, doctors are rightly cautious of using morphine as it should be used. And in consequence there is a vast amount of quite unnecessary pain. This pain is not due to the fact that we have a choice between relieving pain and ruining character. It is just one result of an immoral economic system.

STIMULANTS

Of the various drugs which act on the nervous system, some are often called stimulants. This is almost always incorrect. Alcohol makes you think you are working better. But tests show the contrary. Like cocaine and other drugs which cause excitement, it first paralyses those parts of the brain which have been evolved in the last 60 million years, since the time when most of England was under a shallow sea where the chalk was formed.

The older parts of the brain are concerned with primitive activities, such as eating, fighting and mating, and the newer parts are partly concerned in controlling these older parts, so that if we do these things we do them at the right time and place. They are also concerned with the fine adjustment of muscular movements which can be carried out in a rough way by the older parts of the brain.

So the main effect of so-called stimulant drugs is generally to remove inhibitions, to give the older parts of the brain a holiday, so to speak. This may sometimes be worth doing, but it is certainly not worth doing as a habit. And it can be done without drugs or alcohol, as is shown by one of the many stories about Willie Gallacher in Moscow. The door-keeper of the block of flats where he lived complained that he and some other Scottish comrades had been engaged in a drunken orgy. It was conclusively proved that they had had nothing stronger than buns and milk. But their minds were sufficiently young and healthy to allow them to enjoy themselves without alcohol.

The effects of want of oxygen are extraordinarily like those of alcohol. You have only to reduce the air pressure to about half, to get the symptoms of drunkenness. I think I am making very good jokes. But I notice other people becoming quarrelsome. And when I come back to normal pressure I have a headache for a short time.

Perhaps the most genuine nerve stimulant has recently been discovered. During hard work and various kinds of excitement, the adrenal glands, which lie above the kidneys, pour into the blood a substance called "adrenaline," which makes the heart work more vigorously, and raises the blood pressure by causing small arteries to contract. It is of no use as a drug to cause general stimulation, because the body destroys it in a few minutes. Otherwise we could not relax after violent exercise.

Now, a few years ago a Chinese doctor called Chen started to investigate the traditional drugs of his country. Some of them were already known in Europe. Others were fantastic remedies, like powdered sharks' teeth. But some were really valuable. A root called *Ma huang* was found to contain a crystalline substance called "ephedrine." A solution of this dropped into the nose will dry up a running cold by making the blood-vessels in the inflamed membrane close. But it also makes your heart thump and keeps you awake, and may make your nose sore. Moreover, its effects do not last very long.

It was found to be very similar chemically to adrenaline, but very much more slowly destroyed. So its effects last for hours instead of minutes. A similar synthetic drug, benzedrine, is even more effective and has a relatively greater effect on the brain. If I have to drive a car for more than eight hours at a stretch, I always take it. And I can take it safely, because I have a low blood pressure normally. But if I had a high one, I might burst a blood vessel in my brain, and fall down with a stroke of paralysis, if I raised it still higher with benzedrine.

Naturally enough, benzedrine is open to abuse. You can still buy it at any druggist's, but it has already made some people very ill, and when it kills one or two it will be put on the dangerous drugs list.[1] It does not seem to impair thought or muscular skill like other so-called stimulants, though, of course, it leaves you very tired the next day.

Another traditional remedy which has proved very valuable is digitalis. In the eighteenth century William Withering, a Birmingham doctor, found that "wise women" in the villages used foxglove leaves to cure dropsy. He found that the cures were genuine, though we now know that its main action is on the heart (and that dropsy is only cured when it follows heart trouble).

The greatest use of digitalis is in one group of heart diseases in which the pulse is fast and irregular. The ventricles of the heart, which do the main business of pumping, normally receive just over one impulse per second from the auricles, which are subsidiary pumps. In some forms of heart disease the auricles over-drive the ventricles. In consequence, they have no time to relax between beats, and work inefficiently. Digitalis blocks the process by which the ventricles are stimulated, and they start working at their natural, rather slow rhythm. The pulse rate may fall from 150 beats per minute to 50 within a few hours, and the heart actually pumps more blood at this slow rate than at a fast one, as a worker often produces more during an eight-hour than a twelve-hour day.

Such are a few examples of drug action, facts which would be common knowledge if we spent as much time reading biology as we do in reading advertisements of drugs which are often fraudulent. Some beautiful examples of the latter are to be found in *Fact* for June, 1938, though

[1] This was done in January, 1939.

unfortunately the law of libel prevents the publication of names.

Meanwhile workers can economise very greatly by buying their drugs from the chemist, and not paying the huge surplus value which goes in profits and in payment for advertisements. Here is a list of simple remedies with their actual retail prices, and doses for an adult. Children should take half or less.

Saline Laxatives.—Cream of tartar or sodium phosphate. Each costs 2*d.* per ounce. The dose is about 4 grams or 60 grains, so a dose costs a farthing. Some people will prefer the taste of one and some of the other. They are best taken dissolved in water. Milk hides the taste of sodium phosphate from children.

For Headaches, Neuralgia, etc.—Acetyl-salicylic acid (one trade name is aspirin). Cost 6*d.* per ounce. Dose, ½–1 gram or 7–15 grains. Some people's stomachs are upset by it.

For "Nerves."—Sodium bromide. Cost 6*d.* per ounce. Dose, 1 gram or 15 grains several times daily. But do not continue for more than a few days.

For Sleeplessness.—Paraldehyde. Cost 4*d.* per ounce. Dose, about 4 cubic centimetres (60 minims). Take it with cracked ice or ice water.

You will need a small balance or measuring glass. But as these things cost from five to ten times as much in the form of proprietary articles, you will get your money back in a few weeks, and you will learn something about the economics of the drug trade.

BACK TO NATURE?

Among the numerous critical letters which I have received as a result of my articles on drugs have been several from readers who believed that while sodium bromide and paraldehyde, for example, were "vile chemicals," herbal preparations were "natural," and therefore superior to artificial ones. This cannot be a universal rule, for opium and hashish are prepared directly from herbs, and morphine, cocaine, strychnine and many other deadly substances are extracted from them.

Nevertheless, there is something in the argument. We have got much farther from Nature than was necessary. We meet just the same problem in politics. The history of the Labour movement is full of attempts to go back to primitive communism. But they failed, and we now know that communism can only succeed if we take over the machines developed under capitalism, but see to it that they are our slaves, instead of the other way round.

The story of man is largely the story of substitutes for natural products. We do not know whether primitive men lost their body hair because they started wearing clothes or started wearing clothes because they had lost their hair. But the latter is more likely. The result is that man can live in a greater range of climates than any other animal, provided he can dress properly. But we easily become the slaves of clothes, like the unfortunate guardsmen who sweat under bearskins in summer.

Again we are apt to forget that our domestic animals and plants are even farther from Nature than ourselves.

Wheat is a wonderful improvement on natural grasses from the human point of view. But if it were not artificially sown, and were left to compete with weeds, I doubt if there would be a single wheat plant alive in England twenty years hence.

As for our domestic animals, some would survive, but they would change very rapidly. Captain Cook landed a few domestic pigs in New Zealand in the late eighteenth century. Within 100 years their descendants had changed into lean animals with formidable tusks, although they had an easier time than most animals, as there were no wild beasts to hunt them.

And just as we have transformed wheat and pigs from their original types, we have certainly transformed ourselves, though indirectly, by transforming our surroundings. This has been strikingly true as regards disease resistance. Perhaps civilised men are less tough than primitive as regards resistance to heat, cold, damp and starvation. But primitive men die like flies in the autumn when they are exposed to diseases which are favoured by overcrowding—such as measles—which do us little harm.

The right question to ask about any divergences from Nature is whether they do more harm than good. Are houses harmful? Almost certainly not. But too great a concentration of population is so. Even the well-to-do people in the towns have shorter lives on the average than agricultural labourers.

Are artificial foods harmful? Those who say "Yes" are apt to forget that bread is extremely artificial, and was probably invented less than 10,000 years ago. Butter, again, is no more a natural product than margarine. We have to investigate each food in detail. If we do we shall certainly find that many of the recently introduced foods are less nutritious than the older ones. But some are not. A margarine properly equipped with vitamins is almost certainly better than butter from a cow which has spent the whole winter in a shed living mainly on oilcakes.

It is the same with drugs. Very few are invariably harmful. All are harmful if we take too much of them. This is as true for natural as artificial products. To go "back to Nature" here generally means to take substances in unknown amounts instead of known ones. This is all that happens if, for example, we take foxglove tea, instead of a tested preparation of digitalis, for heart trouble.

I am sure that men, women and children should not be allowed to eat or drink new chemicals until they have been very thoroughly tried out on animals. This is generally done with new drugs, but not always with colouring matters in food or substances used for flavouring sweets.

Two winters ago the S. G. Massengill Company of Bristol, Tennessee, U.S.A., sold a preparation of a quite useful drug dissolved in ethylene glycol, a rather virulent poison. Up to May of last year they had killed seventy-six people and made many others ill. The company said they did not know ethylene glycol was a poison. So, as the stuff had been sold in the sacred cause of profit, no one was electrocuted or even imprisoned. The gangsters must have felt very sore about it, considering the fuss made when one of them bumps off even half a dozen people.

If every new invention which takes us farther from Nature were judged on a basis of social utility rather than individual profit or prestige, I do not doubt that we should reject a great many artificialities, including stiff collars, bombing aeroplanes and high-speed motor cars. But we should realise that a complete return to Nature would mean living without clothes, houses, cookery or literature.

All such slogans as "back to Nature" are meaningless unless we consider the economic system within which the change is to operate, and very often, as in this case, we find that within a better economic system the change would be largely unnecessary.

HEREDITY

SOME FALLACIES

THE study of heredity is of the greatest importance for a number of reasons. It is applied to agricultural plants and animals. It has a certain practical application to social matters. For we do not want children to be born destined to be blind, deaf, or mentally defective, and we should like them to be born with a hereditary bias towards health, intelligence and other good qualities.

It is also theoretically important. The Nazis believe that some races are superior to others, that a race loses its good qualities by mixing with others, and that it can be improved by wholesale sterilisation of the "unfit." Some eugenists in Britain think that the poor are congenitally inferior to the rich, and should be discouraged from breeding.

On the other hand, Marx and Engels accepted Darwin's theory of evolution, not indeed as exactly true, but as the first sketch of a true theory. And Darwinism is based on the theory of "the preservation of favoured races in the struggle for life." If we do not read Darwin's books in detail, we may easily interpret this phrase as justifying the Nazi theory.

But though Darwin sometimes wrote of "lower races,"

he certainly did not regard Europeans as necessarily superior to Negroes. Here is what he wrote to his sister from South America in 1832, in the course of an attack on slavery: "It is impossible to see a negro and not feel kindly towards him; such cheerful, open, honest expressions, and such fine muscular bodies. I never saw any of the diminutive Portuguese, with their murderous countenances, without almost wishing for Brazil to follow the example of Hayti,[1] and indeed considering the enormous healthy-looking black populations it will be wonderful if, at some future day, it does not take place."

Of course, we have learned an immense amount about heredity since Darwin's time, and indeed we must begin its study with facts which he did not know. If we compare two different animals, say two dogs, some of the differences between them are probably due to heredity. One may be a greyhound and the other a fox-terrier. Others are due to environment. One may have a long tail, while the other has had its tail cut off.

But other characters are not so obviously due to one or the other. Black and brown puppies may appear in the same litter. Since they had the same parents, one might be disposed to put the difference down to a difference of environment before birth. But this is not so. Again some people think that differences originally due to environment are later handed on by heredity. This is also untrue in the vast majority of cases.

A great many experiments have been made to demonstrate the transmission of acquired characters, and almost

[1] A former French colony, which achieved independence, and has been a Negro republic since 1804.

if not quite all have failed. It is obvious that mutilations are not inherited, or Europe would be full of crippled children of ex-soldiers who lost their limbs. And the same is true even when animals are mutilated for many generations, as in the case of dogs whose tails are cut. But, of course, some diseases such as syphilis can be handed down from parent to child. This is not really heredity, but infection.

Nevertheless, many breeders think that if you fatten up pigs for some generations, their descendants will tend to become fat, and so on. Similarly some plant-breeders think that if seeds are induced by altered conditions, such as Lysenko's technique of vernalisation, to germinate earlier than usual, this habit will be inherited. Others think that habits, such as preferences for food, can be handed down.

In such cases there has generally been selection from a mixed lot. The pigs which fatten up best are chosen as ancestors for future generations, the seeds which germinate earliest yield the biggest plants and the most seed for the next generation. In fact, the change of environment has had an effect, but an indirect one.

Experimenters can get over this difficulty by making what is called a pure line by prolonged self-fertilisation or inbreeding. This may lead to sterility or early death; but in many plants, such as wheat and peas, and some animals, such as mice and guinea-pigs, it does not. The members of a pure line are not all alike, but the differences between them are not inherited. The children of the lightest guinea-pig in a pure line are no lighter in colour than those of the darkest. Such differences are merely temporary effects of environment. Careful experiments show that new habits are not inherited. Payne bred flies of a species accustomed to fly towards light in complete darkness for sixty-nine generations. He found their descendants no more and no less attracted by light than abnormal flies.

Reactionary biologists, such as Professor Macbride, who thinks that the unemployed should be sterilized, naturally use the theory of the transmission of acquired habits for political ends. It is silly, they say, to expect the children of manual workers to take up book-learning, or those of long-oppressed races to govern themselves. Laboratory experiments agree with social experience in proving that this theory is false.

THE PHYSICAL BASIS

If we are to understand the facts of heredity, we must first know a little about their physical basis. Living beings can reproduce in two ways, with and without sexual union. The second sort of reproduction, which we find in plants which can be propagated by cuttings, and in insects such as greenfly, where the female generally reproduces her kind without a male, almost always gives progeny very like the mother.

The first kind involves the union of two cells. The mother produces an egg cell, which may be large, like a bird's egg, barely visible with the naked eye, like a rabbit's or woman's egg, or only visible with a miscroscope, like some insects' eggs. The father produces a cell which is almost always too small to see, a swimming spermatozoön in animals, a pollen grain in plants. These unite, and the new cell grows and divides, until we may get a large plant or animal made of millions of millions of cells.

Thus in each generation the stream of life goes back about 1,000 million years to the stage of the single-celled animals and plants from which we are descended. A thousand millions years of evolution are negated. But then follows the negation of the negation. The single cell becomes a many-celled animal or plant. Every Marxist will expect to find novelty arising. And so it does. If you divide a geranium plant or a potato, or take a graft from an apple tree, you get plants like the original. If you grow its seeds, you get something rather different from the parent, generally much less useful to man.

Since the egg is so much larger than the pollen grain or sperm, one might expect, if heredity has a material basis, that the offspring would always resemble their mother more than their father. This is not so in most cases. The material basis of heredity is found in a part of the cell called the "nucleus" which is no bigger in the egg than the sperm.

When the cell divides into two, for example, in the human skin where this is constantly happening to replace losses, the nucleus divides first, and we can see with a microscope that it is organised into a definite number of tiny threads called "chromosomes," forty-eight in a man or woman, forty-two in a wheat plant, twenty-six in a frog, fourteen in a pea, and so on. Each chromosome divides in two, so the new cell has the same number.

But when eggs or male sexual cells are being formed, the number is halved, and made up again when the two cells join up. Thus, of a woman's forty-eight chromosomes, twenty-four come from her father and twenty-four from her mother. The two sets are so like that we cannot tell them apart with a microscope, but they may carry very different "genes," as the units concerned in heredity are called.

Let us see what a gene means. Every ordinary person, at a particular point in one pair of chromosomes, has a gene concerned in making his fingers grow. But one family in North Wales includes a number of people with short fingers, though quite healthy. They have got a gene for finger growth from one parent, but in the set of chromosomes derived from the other parent this gene is absent, or at least not active.

One gene for finger growth is nearly as good as two, but not quite. So their fingers are rather short. If such a person has children by a normal husband or wife, all the children will get a gene for finger growth from the normal parent. Half of them, on the average, get a normal gene for the short-fingered parent, and have normal fingers. But half of them get a growth gene from one parent only, so their fingers are short.

Thus a short-fingered person hands on the defect to about half his children, and it never skips a generation. The abnormality has been handed down for a long time, and there is another branch of this Welsh family in the United States.

If, however, two short-fingered people marry, things are more serious. One in four of their children get two genes for finger growth, and are quite normal. One-half get one gene, and are short-fingered. But one in four gets no genes for finger growth at all, and is born a hopeless cripple with no fingers or toes, and other deformities.

Quite a lot of human defects are handed down in this way—that is to say, never skipping a generation, and going to half the children. Several kinds of paralysis developing in middle life, and some kinds of blindness, are examples, though we rarely know what happens if two abnormal people marry. It is to people with such defects as these that the slogan "Sterlise the unfit" could best be applied. But most congenital ailments are inherited in other ways of which I will write later.

And even in these cases, things are not always simple. Cataract is a disease where the lens of the eye becomes opaque, and it may cause fairly complete blindness. Many old people get it, but where it develops in childhood it is usually inherited in this way. In a family with cataract we often find that the degree varies. Thus in one Dorsetshire family, a woman went blind with it. She gave it to one of her sons, but it was very mild, merely producing specks such as many people can see if they look at a clear sky or a white wall.

But one of this man's sons had it so badly that he had to be operated on as a baby. If eugenists say that this man should have been sterilised, they are advocating the sterilisation of the fit. In any case, there are two alternatives to sterilisation—namely, celibacy and birth control. It is typical of the outlook of many eugenists that they do not suggest these remedies. But many sufferers from congenital diseases of this kind voluntarily avoid having children. And in a properly educated community, I believe that almost all of the few hundred people per million with serious transmissable defects would avoid parenthood.

WHY MARRIAGE IS A LOTTERY

In the second article of this series I described what is meant by the word "gene"—namely, something in the nucleus of a cell which influences development, and is reproduced when the cell divides, and handed down to about half the children. This simple principle enables one to understand and control animal breeding to a very great extent.

Let us take an example from rabbit breeding. The ordinary rabbit has black hairs with yellow bands on them. The "chinchilla" rabbit has black with white bands, giving a much clearer grey. If we cross a pure-breed rabbit of the wild colour with a chinchilla, the first generation get a gene for making yellow pigment from the wild-coloured parent. One such gene per cell will make as much yellow pigment as two, so the hybrids are yellow-grey like the wild rabbit.

But if we cross these hybrids to chinchilla, only half the children get a gene for making yellow. So half the young are wild-coloured and half chinchilla. If we cross the hybrids together three-quarters of the young, on the average, get a gene or two genes for making yellow, and only one-quarter get a chinchilla gene from each parent. So three-quarters of the young are wild-coloured and one-quarter chinchilla.

There is another variety of rabbit called "rex" or "castorrex" which has none of the long coarse hairs of ordinary rabbits, but only the soft velvety under-fur. When crossed with the ordinary rabbit this character behaves just like chinchilla, disappearing in the first generation, and reappearing in one quarter of the second if the hybrids are bred together.

In order to combine these characters, chinchilla and rex rabbits were crossed together. The hybrids had coats of the colour and texture of wild rabbits. But on crossing them together, one quarter of the young were short-furred rex, and of these again one-quarter—that is to say, one-sixteenth of the total number—were chinchilla. The chinchilla rex were then bred together and a new breed established. Their skins make a very pretty fur.

Of course, this is a very simple case. Only two genes were concerned. Whereas the differences between two breeds are generally much more complicated. And where they differ by ten or twelve genes, as, say, a Sebright bantam differs from a Rhode Island red fowl, you would have to breed a second generation of several millions to get all the possible types.

How does this apply to human beings? The key principle is that a normal-looking man or woman, like the normal-looking hybrid rabbits, may carry, alongside a gene for a normal character, a hidden or "recessive" gene for an abnormal one. The abnormality may be fairly harmless—for example, colour-blindness or albinism—but it may be very serious indeed—for example, one of the many kinds of blindness or idiocy.

Probably most of us carry a hidden gene for some defect. But no harm comes unless we happen to marry someone with a hidden gene of the same kind. Unfortunately, there is no way of detecting these hidden genes in most cases, which is one reason why marriage is a lottery. But if we marry a near relative there is a good chance that he or she will carry the same hidden gene. For example, if I have a hidden gene for deaf-mutism the chances are several hundreds to one against my marrying a woman with the same gene, in which case some of our children may be born deaf.

But if I marry my first cousin, there is a chance of one in eight that she will have the same hidden gene. And if I were allowed to marry my sister the chance would be one in two. So a great many congenital diseases are very much commoner when the parents are related than when they are not. Dr. Usher, a London eye specialist, examined the family history of forty-one patients with retinitis pigmentosa, a disease of the eyes which causes partial or complete blindness, and sometimes goes with deafness or mental defect.

In seven cases the parents were first cousins, and in another first cousins once removed. Now, only about one person in 150 in London marries a first cousin. So this means that if you marry your first cousin you are about twenty-five times as likely to have a child with this disease as if you married an unrelated person. If you married your sister, you would be 100 times as likely.

So there is a very good biological reason for making incest a crime; and some reason for discouraging cousin marriage. Though it must be remembered that most of the children of first cousins are healthy enough. Some congenital diseases are commoner among Jews than the rest of the population. This is not because the Jews are a degenerate race, but because in countries where they are cut off from the rest of the population by persecution or by their own prejudices they often marry near relatives.

It is quite true that many fine animal and plant races have been built up by inbreeding. And some eugenists would like to apply the same methods to men and women. But inbreeding quite often produces abnormalities in animals as well as men. The animal breeder wrings their necks, and only breeds from his best specimens. As we do not kill human beings with physical and mental defects we had better not start on a system of breeding which would produce them.

All the facts mentioned in this article were discovered since 1900, though the clue was given by Mendel, a Czech abbot, in 1865. Why were they not discovered sooner, since they can be discovered without any apparatus? Very likely, facts of this sort were known to shepherds and herdsmen in the old days. But this knowledge was never written down, and must have been lost as breeds became more and more standardised.

When learned men took to studying biology, bigger profits could be made by the crude exploitation of colonies than by improving livestock. So their first efforts were directed to classifying plants and animals. Only at the end of the nineteenth century was it necessary to make new wheats for Canada, instead of using it as a source of furs. As soon as the economic demand arose, the laws of heredity in plants were studied. And these laws turned out to be applicable to animals, and among other animals to man.

EXCEPTIONS TO THE RULES

ANY Marxist who has read the first three articles in this series must, if he or she understands dialectics, have said something like this. "Isn't Haldane trying to give a mechanical account of heredity in terms of genes? If that were the whole story there could be no real progress, only a rearrangement of genes." This criticism is quite right. The rules which I have given generally work. I might breed 10,000 rabbits without finding an exception.

But the ten-thousandth rabbit may be an exception, perhaps a black when he should be a blue according to the rules, but more likely a weak or deformed animal. Of course, most deformities are due to bad environment before or after birth, and are not inherited. But a few are. Sometimes a new gene suddenly appears, and is afterwards inherited according to the usual laws.

This process is called "mutation." It is an accident in two senses. We cannot prevent it happening, though we can speed it up with X-rays. And it is generally harmful. Most mutations change an animal or plant into one which is less fit than the normal in some way.

But it may be better suited for human purposes. For example, peas with soft pods have arisen by mutation. They are easier for birds and caterpillars to eat, so they would soon have been killed off in the wild state. But they are also easier for men to eat, so we have bred them.

Wild rabbits would be hampered by long hair, but the long-haired variety arose by mutation and was kept by men, first as a curiosity, and later as a source of wool. However, many other mutations have given forms which are useless both in Nature and domestication—for example, peas that grow into plants with no tendrils, and rabbits with no fur. These useless mutations are weeded out by natural selection and by human breeders. And, in fact, natural selection is mainly concerned with weeding them out. Only about one mutation in fifty, if so many, seems to be of any value to the species, and spreads as the result of natural selection, so that the species evolves.

Now we can see why we must be very careful in arguing from the results of animal breeding to human inheritance. It is not because the laws are different, but for the following reason. Breeds of domestic animals have been established as the result of mutation. A hornless cow or a long-haired rabbit suddenly appears. Great numbers of their descendants are bred, and perhaps all the Angora rabbits and all the Polled Angus cattle in Britain are descended from a single original animal which showed the new character.

So we can predict that the result will be much the same whenever we cross a long-haired and short-haired rabbit.

But this is not so in man. More often than not, the difference between brown and blue eyes is due to a gene of which brown-eyed people have one or two and blue-eyed people none.

If this were so, two brown-eyed parents could sometimes have a blue-eyed child, but two blue-eyed parents could never have a brown-eyed child. This certainly does not happen very often, but it happens often enough to be quite inexplicable by illegitimacy. And no wonder. Primitive men were probably brown-eyed, and there were several different mutations which gave rise to blue eyes. So the inheritance of eye colour is fairly complicated. Unfortunately, the statement that blue-eyed parents cannot have brown-eyed children is still made by some writers and lecturers, and gives rise to quite unjustified suspicions.

With our new knowledge about mutation, we can bring Darwinism up to date. If you try to make a small group of animals or plants evolve by selecting the heaviest or hairiest as parents, you generally get success for the first ten generations or so, and then selection makes no further progress. You have got together all the genes making for weight, hairiness or whatever you are breeding for, and you can do nothing more until a new gene working in the desired direction turns up by mutation.

If this happens once in 10,000 animals, and you are working with a stock of 100, you may have to wait 100 generations. But in a natural population of 1 million, there would be 100 mutations in each generation and progress would be steady. The change of quantity becomes a change of quality. We can now view evolution, not as an effect of mutation or selection alone, but as an effect of the struggle between them.

If we had mutation alone, every species of animals and plants would degenerate into a collection of freaks, with perhaps one in a million fitter than the original type. If we had selection alone, evolution would come to an end fairly quickly. Now, at first sight the main effect of natural selection is merely to weed out harmful genes which arise by mutation, such as those giving rise to toothless mice or men whose blood will not clot. And until I studied Marxism I wrote of the balance between the two.

But really the balance is probably never complete. Sometimes evolution is very quick. For example, a new gene arose near Manchester about 1848 in the peppered moth, which turns it black. These blacks seem to be fitter than the normal type in industrial areas, perhaps because they are less easily seen against a black background. By 1900 the black variety had completely replaced the old type in most English industrial districts, and also in the Ruhr district of Germany.

Such speedy evolution is rare, and we are only just starting to collect evidence of slower evolution, which can only be done if careful censuses and measurements of animals or plants are made. Work of this kind is very tedious, but it is absolutely necessary if we are to understand the principles of change in Nature, which turns out to be often due to a struggle between opposing tendencies which at first sight balance one another.

CATS, KINGS AND COCKERELS

It is not quite true that the parents are equally important as regards heredity, even when we have allowed for the fact that the mother provides the environment for the first nine months of human life. There are many characters where the father has no influence at all on his sons, though a great deal on his daughters. One of the simplest of these is found in cats. Some cats are black, with or without a tabby pattern. Others are yellow. Both kinds breed true, though they may give dilute colours, such as blue due to a hidden gene, or white-spotted kittens. And no matter whether we mate her to a black or a yellow male, a yellow female will only give yellow sons, and a black female only black sons. The father makes no difference.

But if we cross black and yellow either way, the daughters are all tortoiseshell, that is to say black and yellow or tabby and yellow. The reason is as follows. A female cat (or a woman) has two chromosomes called X in each cell, a tom-cat (or a man) has one X with a smaller mate called a Y. Every egg gets one X. The sperms may get an X or a Y. The sperms carrying X produce females, those carrying a Y produce males.

Now, the genes for black and yellow are carried in the X, but not in the Y. So a tom-cat can only be black or yellow; a female can get both genes and be a tortoiseshell. We can now work out how the tortoiseshell cat should behave. Her sons get black and yellow genes in equal numbers. Mated with a black tom, all her daughters get black from the father, and half get it from the mother, the other half getting yellow. So half the daughters are black and half tortoiseshell. Similarly a tortoiseshell female and a yellow male give tortoiseshell and yellow daughters.

The human X chromosome carries a gene needed for normal colour vision. This has gone wrong in about 2½ per cent. of cases. So 2½ per cent. of men are colour-blind. However, one gene for normal vision will generally do the work of two, so very few women are colour-blind. To be so they must have got a gene for colour-blindness from both parents.

A much more serious disease is hæmophilia, in which the blood does not clot for some hours after shedding. The normal X chromosome carries a gene which makes a substance concerned in clotting. This gene is inactive in a hæmophilic. If a hæmophilic marries, unless he marries into a hæmophilic family, all his children are normal. His sons get a normal X from their mother, and do not hand the disease on. But his daughters have one normal and one abnormal X, so half their sons are hæmophilic.

There are several degrees of this disease. But the commonest type is pretty fatal. The patients often bleed to death as boys, and if they survive to manhood, are frequently crippled by bleeding into the joints. Why, then, does not the condition disappear by natural selection? Because about as many new hæmophilic genes appear in each generation by mutation as are wiped out by natural selection. About one gene in 100,000 for normal blood-clotting changes over to a gene for hæmophilia.

Curiously enough, this happened in Queen Victoria, or one of her parents. As a result of mutation, she became a carrier for hæmophilia. One of her sons, Prince Leopold, was hæmophilic, but King Edward VII was not, so there is no hæmophilia, either open or hidden, in the present British Royal Family. But two of her daughters got the gene for hæmophilia. Like their mother, their blood clotted normally, because they had got one normal X from their father.

But one of them bore one hæmophilic son, the other two. However, it was their daughters, the granddaughters of Queen Victoria, who made history. For one married the last Tsar of Russia, the other the last King of Spain. And in each case the eldest son was a hæmophilic. Possibly in Spain, and certainly in Russia, this fact helped the Revolution. For it was by his claim to be able to stop the Tsarevitch's bleeding that Rasputin kept his hold over the Tsar and Tsarina. And Rasputin was one of the influences which broke up the unity of the Russian ruling class, and made the Revolution of March, 1917, easier.

So Queen Victoria did her bit for revolution, though unconsciously! We cannot put the hæmophilia in either the Russian or Spanish royal families down to mere bad luck. In each case the ruler married the sister of a hæmophilic, that is to say put "high birth," or in other words snobbery, before health.

Characters inherited in this way are called sex-linked. In birds they are used for a different purpose. Here the male has two X chromosomes, and the female one. Now the light Sussex breed, which are white with a few black feathers, have a gene in their X chromosomes which stops the formation of yellow pigment in the feathers. So if we mate a Rhode Island red hen with a light Sussex cock, all the chickens get this gene, and are white.

But if we mate a light Sussex hen with a Rhode Island red cock, the cockerels get this gene, and are white, but the pullets do not get it, so they are yellow. Hence the sexes can be distinguished at hatching, which is extremely difficult otherwise. And the young cockerels can be at once fattened up for eating, while the pullets are kept as egg-layers. This is such an advantage that in England in 1935 one single firm reared 800,000 chicks from sex-linked crosses of this kind, and several million were reared in the whole country. Several other genes in poultry behave in the same way, and so do genes in ducks and canaries.

As a man, I wish I could point to some good characters which are inherited in this way, and are therefore commoner among men than women, like hæmophilia and colour-blindness. But unfortunately none are known. And besides these diseases, quite a number of others, including some forms of blindness, are due to genes in the X chromosome. This may be one of the reasons why women, on the average, live longer than men.

MENTAL DEFICIENCY

About one person in 200 in England is certified as a mental defective or sent to a special school, and a good many more could be certified under the existing law. A few of these people are imbeciles who are incapable of doing any but the most mechanical work, and idiots who cannot even speak or dress themselves. But many of them, though rather slow and stupid, can learn a trade, and would be useful members of a society where there was work for all.

Some children are labelled as defectives merely because they cannot learn to read. This is certainly a handicap, but such people may be very able in other ways. Claude Gelée, a great seventeenth-century French painter, was a "defective" of this kind. However, when all allowances are made, there are plenty of genuine defectives.

Now some people say that if everyone was put in a proper environment there would be no mental defect. Others put it down to heredity, and say that if all mental defectives were stopped from breeding the condition could be almost wiped out. The most scientific study of the question so far made is that of Dr. Penrose, who examined 1,280 defectives in an institution at Colchester, and collected information about 29,000 of their relatives.

The first point which comes out of his work is that it is as silly to talk about the cause of mental defect as about the cause of blindness. The eye can go wrong for hundreds of different reasons, and the brain, which is much more complicated than the eye, for thousands. Some of the mental defect was caused by injury at birth or blows on the head in early childhood. These cases could have been largely prevented by proper care. Others were due to congenital syphilis—that is to say, infection before birth. This, again, can be prevented by medical treatment of the mother. Still others were caused by childhood illnesses, such as meningitis, which are harder to prevent. There were several cretins—that is to say, dwarfs who would have been nearly if not quite normal if their condition had been found out soon after birth and treated by feeding an extract of the thyroid gland.

Others, again, were emotionally abnormal. They had been unable to learn at school because of abnormal feelings rather than weak intellect. Many of these children would have been normal in better homes, and many could have been cured by psychological methods as children. Some, on the other hand, would probably have become defectives however well they were cared for.

Still others suffered from a disease called "mongoloid imbecility." This is due to bad conditions before birth, and is mostly found in children of old mothers. The average age of mothers of those children is about forty. If economic conditions allowed early marriage we could have all the children needed to keep up the population, and yet far fewer of these defectives would be born. By postponing marriages, the Means Test not only causes unhappiness, but actual mental defect.

However, all these cases only make up about a quarter of the total. What about the rest? Are they due to heredity? Certainly not in the ordinary sense of the word. Only about 8 per cent. of these defectives had a mentally defective parent. However, in many other cases heredity is at work. Epilepsy is sometimes inherited, though it may be due to injury. And the child of an epileptic who is not feeble-minded may be so itself. There are also several diseases due to single genes which cause the formation of tumours. These may be in the skin, heart and other organs, and if so they do not affect the mind. But in the brain they may cause mental defect. The few thousand people who suffer from these diseases certainly should not have children.

A good many other cases are due to hidden genes. Both parents may be normal, but carry a gene for a small head or some other abnormality. And if a child gets such a gene from both parents, it will be defective. None of the parents of the small-headed idiots who made up 2 per cent. of Penrose's cases had small heads themselves or were mental defectives. Until we can discover a method of detecting hidden genes, there is no way of stopping the birth of such children, though a few less would be born if the marriage of cousins were discouraged.

Thus sterilisation would not go far to prevent the birth of defectives. If every defective in the country had been sterilised in the last generation, only 8 per cent. of Penrose's patients would not have been born. Even if all dull and backward people had undergone this fate, three-quarters would have remained.

The demand for sterilisation is based mainly on economic, not on biological grounds. I quite agree that genuine mental defectives should not breed. Whether or not they transmit their defect, they cannot give their children a satisfactory home. Most of them are far happier in institutions than outside. But they cost about £1 per week each, roughly half of which comes out of taxes. It is argued that this money could be saved if they were sterilised and then turned loose.

Now if there was work for everyone in Britain, this would be a fairly good argument. But as things are, if they get work at all it is unskilled work at very low wages. And they cannot defend themselves against exploitation by employers. So sterilisation of all defectives, as an alternative to proper treatment in institutions, would do only a little to prevent mental defect, and quite as much to increase unemployment and to lower wages.

RACES

THE idea of superior races plays an important part in Nazi propaganda. The Germans have a right to rule others because they are a superior race, and the Jews must be expelled because they are inferior. The same sort of arguments are used by the British in India, and by many of the whites in South Africa and the Southern States of the U.S.A.

To examine these theories, we must first ask what is a race, and whether the Germans, for example, are a race, and then we must ask whether some races are superior to others. We may define a race as a people who normally breed together, and who differ from other peoples as regards inherited physical characters which are found in all members of the race.

No race is known whose members are all exactly alike, and the differences within a race are partly inherited. At first sight, it is easy to pick out a number of races—for example, Europeans, Chinese, Negroes, Red Indians,

Australian blacks, and so on. But things are not so simple. If one went overland from Sweden through the Soviet Union, Turkey, Syria, Palestine and Egypt to the Sudan, one would find skins gradually getting darker, but nowhere could one draw a sharp line and say that everyone north of it was blacker than everyone south of it.

Of course, a Nazi might say that this was due to an admixture of originally pure races. There is no evidence for this. We don't know what men's skin colour was like 10,000 years ago, but we do know what their skulls were like. And we find that the skulls of people living together in those days varied just as much in shape as they do now. Probably race mixture and race formation have been going on together as long as man existed. And mankind is probably very much better for it. If there were no race differences, the world would be a much duller place. If there were no race mixture, there might be several different human species incapable of breeding together, and the Nazi doctrine would be true.

There have been men in the Old World for hundreds of thousands of years, and they have developed fair races in northern Europe, and also several different black races in the tropics—namely, Negroes in Africa, Dravidians in Southern India, Papuans in New Guinea, and so on. But men only came to America from Asia about 10,000 years ago, so they have had no time to develop a black race in Brazil, though the Brazilian natives are mostly darker than the Red Indians or Eskimos.

We can certainly say that Englishmen and West African Negroes are different races, in the sense that you can always tell an Englishman from a Negro. But you cannot draw such sharp distinctions within Europe. Most Swedes are fairer than most Spaniards, but there is some overlap, and the darkest Swedes are darker than the lightest Spaniards. However, within Europe there are several different physical types.

The most important are the Nordics, tall, fair, and long-headed; the East-Baltics, tall, fair, and square-cheeked, a common type in Russia; the Alpines, brown-haired and round-headed; and the Mediterraneans, dark, short, and long-headed. But nowhere can you find a people consisting entirely of one type. There is no Nordic race, as there is a Negro race.

Some European countries have a fairly uniform physical type. But unfortunately for the Nazis, Germany is not one of them. There are Nordics in the north-west, East-Baltics in the north-east and Alpines in the south, particularly in Bavaria and Austria. After Germany, Italy is perhaps the most racially mixed of the great European nations. Of course, the Soviet Union is still more so, but the different nations composing it are mostly fairly "pure," though, of course, members of all of them are found in Moscow.

So if Europe were divided upon a basis of race—that is to say, of innate physical characters—Germany would be split up, some parts being united with Poland, other with Holland, Scandinavia, Switzerland and so on. As for the German Jews, they are on the average more Asiatic in their physical characters than the other Germans in the west, but much less so than the East Prussians.

I believe in the superiority of some races in one respect: Europeans are on the whole superior to Negroes in a cold climate, because they are better adapted to it. But the Negro, with his dark skin to protect him from sunburn, his extra sweat glands, and his immunity to yellow fever, is superior to the European in West Africa. As for intelligence, it is certain that races overlap, for clever Negroes are cleverer than stupid Englishmen, and musical Englishmen are more musical than unmusical Negroes.

We don't know much about averages. In the United States whites do better than Negroes, on the average, in intelligence tests. But this may have nothing to do with race, for education counts in these tests. In the Army tests of 1917, the Negroes of Ohio scored a higher average than the whites of Arkansas. Even if it were found that given opportunities, whites do better than Negroes on tests drawn up by whites, it is quite likely that Negro examiners could design tests on which their own race could beat the whites! Questions of this sort will perhaps be decided in the Soviet Union when the different races have enjoyed real equality for another generation.

To a biologist, one of the most striking arguments against colour prejudice is furnished by thoroughbred horses. Many animal breeds are alike in colour. But the race-horse is selected on the ground of performance, and race-horses are of many colours, mostly bays, browns and blacks, but including even a few greys, such as Squadron Castle, who won the Lincoln Handicap this year.[1] And a team to represent the human race, whether of Olympic winners or Nobel Prizemen, would include Jesse Owens or Sir Venkata Raman, to mention no other members of "inferior" races.

[1] 1939.

The truth about human races, when we know it, will no doubt be complicated. But one simple theory which is certainly nearer the truth than Hitler's was stated by old Andrew Marvell 270 years ago:

"*The world in all doth but two nations bear,*
The good, the bad, and these mixed everywhere."

SCIENCE AND SOCIETY

MY JOB

IN the middle of the nineteenth century Weldon invented a process, now superseded, for the manufacture of chlorine. Unlike most inventors, he made a fortune, which passed to his son, who became Professor of Zoology at Oxford, and spent much time in measuring animals. For example, he showed that when a breakwater was built at Plymouth, and the water behind it got muddier, the crabs living there became broader, apparently because they needed roomier gill chambers.

He died in 1906, and his widow in 1937. His capital was mostly left to endow a professorship of Biometry at University College, London, and I was chosen as professor. Biometry is defined as "The application of higher mathematics to biological problems." The whole affair is typical of the haphazard way in which scientific research is supported in capitalist countries. Naturally, I should prefer a job with more experiment in it, but I am lucky to have one so near to my desires as this.

I only give forty or fifty lectures per year, and devote the rest of my time to research. I am mainly concerned in applying mathematics to problems of heredity and evolution; and I publish about ten papers a year in various scientific journals. Some of them are quite unintelligible except to specialists, but others can be explained in fairly simple language.

One, which I published this year, dealt with the following problem. Almost all human albinos are the children of normal parents. Now, if two rabbits or rats which are not albinos have any albino children, the proportion of albinos (that is to say, pink-eyed whites) is very close indeed to a quarter, provided a big enough family is bred. And there is a simple theoretical reason for this.

So if the laws of the heredity of albinism are the same in mice and men, we should expect to find three normal children to one albino in the families which contain at least one albino. Actually, the proportion is less than two to one. This is due to a simple fallacy. Most human families are small. So we do not include in our list a number of families which would have included an albino if only there had been more of them.

This kind of pitfall is very common in statistics. If I asked every child leaving school in London this year how many brothers and sisters he or she had, and then calculated the average, it would be much higher than the average family size in London. First of all, I should have no representatives of childless families. Secondly, I should have ten times as big a chance of getting a child from a family of ten as from a family of one. So I should greatly exaggerate the number of large families.

Similarly with albinos. If I examined the families of the half-dozen or so albinos who left school in London this year, I should have three times the chance of hitting on a family with three albino members as on a family with only one. It is possible to make a correction for this fallacy, and one of my papers published this year deals with the correction. Other workers had dealt with the matter before, and I found inaccuracies in some of their work; so I expect the same fate will befall my own. But at least I showed that the corrected proportion of human albinos was pretty close to a quarter. The same is probably true for various kinds of idiocy and other abnormalities.

Another paper dealt with the mating system of beetles. I sometimes go down to the London docks with my colleague, Dr. Philip, and catch large numbers of a beetle which lives in the great bales of sheepskins which are unloaded there from various different countries. These beetles have several different kinds of colours and wing markings whose inheritance is understood.

If the beetles mate at random the various possible kinds will be found in certain proportions. If there is a tendency for like to mate with like or unlike, the proportions will be different. Actually, it turns out that these particular animals mate at random, though this is not so in other species.

In yet a third paper I took up a problem posed by Darwin, and discussed the way in which species with different mating systems may be expected to evolve, comparing, for example, the self-fertilising annual grasses and the outbred perennials. It will take some years to find out whether my theory is true or false. But for any detailed understanding of evolution it must be discussed.

Besides this, I advise my junior colleagues with regard to their work, which is mainly experimental. All this is what is commonly called "pure science," but is really long-range science—that is to say, science which will not find a practical application for some years to come. If we are about to enter a period of declining civilisation it may never be applied, as the journals in which it is published may all be destroyed before men take an interest in such matters again. But I do not doubt that at some future time these apparently rather futile investigations will prove as important as did some of the researches at which Swift laughed in his account of Laputa in *Gulliver's Travels*.

But if, as seems very much more likely, the horrors of our age are not merely the death of capitalism, but the birth of socialism, then my successors in a happier time may think that, allowing for his inevitable *bourgeois* prejudices, old Haldane thought with reasonable clarity on some biological topics. That is the best that I can hope, and it is a very good best.

SOME GREAT SCIENTISTS OF TO-DAY

EVERY year, on November 30th, the Royal Society awards a number of medals to men whom its council believes to have rendered great services to science. The men who got them last year are all but two still hard at work; and their work is a sample of what scientific men think worth while. Of the seven medals given this year, two are Royal Medals given from the public funds for work published in "His Majesty's Dominions"; the other five may be given to foreigners.

The highest award, the Copley Medal, was given to Niels Bohr of Copenhagen, a theoretical physicist. Bohr's theory has been the counterpart of Rutherford's practice. Rutherford was one of the greatest scientists of all time, and there was nothing complicated about his thinking. After he had done an experiment every laboratory assistant could understand it. In fact, when one reads his work one is always saying, "Why didn't someone else think of that?" Probably the unknown inventors who first made simple things like harness, wheels and arches were men of Rutherford's type.

But his work seemed to contradict the previously known laws of physics. Electrons moving round the nucleus of an atom did not behave like electric currents in a wire. Bohr produced a theory of atoms which reconciled the old and the new knowledge. But no one could call it a simple theory. To look at Rutherford, he might have been a New Zealand farmer like his father. Bohr looks like a thinker. When he is thinking, he screws up his face as if he had a severe pain.

Some great thinkers are solitary. Wordsworth described Newton's statue as:

> "*The marble index of a mind for ever*
> *Voyaging through strange seas of thought, alone.*"

Bohr is a very social thinker, and at the Institute for Theoretical Physics in Copenhagen he manages to get his colleagues thinking. As the President of the Royal

Society put it: "He possesses to an extraordinary extent the ability to draw ideas from minds which would otherwise probably never have produced them, and all who have fallen under his influence are conscious of his supreme power of inspiration."

R. W. Wood, of Baltimore, who got the Rumford Medal, is a physicist of a very different kind; an extremely ingenious and skilful experimenter. His work has had many technical applications. For example, his light filters first made photography with ultra-violet and infra-red rays possible. He was the first to make atomic hydrogen in quantity, and this led Langmuir to produce the atomic hydrogen welding torch, which gives the hottest known flame.

Aston, of Cambridge, who was given a Royal Medal, has revolutionised chemistry very largely by sheer manual skill. By an apparatus called the "mass-spectograph," in which electrically charged atoms or molecules are shot through a vacuum whilst being pulled out of their course and brought to a focus by electric and magnetic forces, he has shown that most of the chemical elements are really mixtures, whereas for nearly a century the atoms of one element had been thought to be all alike. Though the mass-spectograph has been in existence for nearly twenty years, its use demands such skill that almost all the important work with it has been done by Aston, and he has also made almost all the improvements in the original apparatus. He is one of the men, not uncommon among skilled workers, who "think with their hands."

Fisher, of London, who got the other Royal Medal,

is a mathematician who has turned to biology. He has been particularly concerned with agricultural experiments; for example, the effects of fertilisers on crop yield, and with heredity. In both these fields statistical methods have become very important, and Fisher has simplified some of them enormously. Many of his biological opinions are very controversial, and I, for one, disagree with some of them. But in trying to disprove them, I at least pay him the compliment of using the mathematical methods which he invented. He is a strong supporter of family allowances.

Barger, of Glasgow, who has unfortunately died since the award, was given the Davy Medal for his chemical researches on drugs, poisons and hormones. Unlike Aston, he did most of his best work in partnership. No man living has done more to clear up the immensely complicated problem of the relation between the chemical constitution of substances and their action on men and animals.

Bower, formerly of Glasgow, was given the Darwin Medal for his work on ferns, especially the ancient ferns whose fossils are found in coal. He is eighty-three years old, and represents the generation of biologists who were mainly concerned in working out the evolutionary pedigrees of plants and animals, whereas to-day the emphasis has largely shifted to a consideration of just how and why they evolve.

Finally, the Hughes Medal was given to Cockcroft and Walton, both young men in their thirties, who were the first to transmute one chemical element into another without using radio-active substances to do so. They are pupils of Rutherford, and designed an apparatus in which protons (the smallest particles with a positive charge) are made to move in an electric field of half a million volts.

There is no doubt that all these men's work could be used for human benefit. It can equally well be used for killing. It is up to the people to decide which way their life's work will be applied.

HOW BRITISH SCIENCE IS ORGANISED

THE British Association was founded in 1831, and at that time almost every serious scientist in Britain belonged to it. There were so few of them that most of the year's work in a given branch of science could be discussed in a few days. In fact it merited the title of "Parliament of Science" which is still bestowed on it by some newspapers.

Since then the situation has completely changed. For example, there is a Physiological Section of the British Association. But so much work is now done on physiology that the Physiological Society, which has nothing to do with the British Association, meets eight times yearly to see new facts demonstrated, and to listen to accounts of new discoveries and discuss them.

There are a number of other societies of the same kind, for example, the Royal Astronomical Society, the Chemical Society, the Genetical Society and the Geological

Society. Unlike the British Association, they are composed of scientists only, and their meetings are rarely reported in the Press.

Finally, there is the Royal Society of London for Improving Natural Knowledge. This has 384 British scientific fellows, forty-nine foreign members, and fifteen British fellows, such as the Duke of Windsor and Earl Baldwin, elected for reasons other than their scientific eminence. Election to it is an honour, though it has its economic advantages, if we can believe the story of an eminent surgeon who, when asked what the letters F.R.S. after his name meant, answered, "Fees Raised Since."

When it was founded nearly 300 years ago, it included every scientist in England, and many others, such as Samuel Pepys, who were interested in science. But now it only includes a small fraction of our scientists, and its discussions are less lively than those of the societies concerned with individual sciences.

On the other hand, the British Association is concerned with matters other than science. It has sections devoted to psychology, which is still only partially scientific, and to education and economics, which in this country at any rate are hardly so at all. So it has fallen away from its former scientific spirit to a certain extent.

The scientific societies generally publish journals in which the results of research appear, and besides these there are other similar journals run by private enterprise. It may take five or ten years before these results are incorporated in books, which are all that most students, or the general public, can read.

But, except for the Royal Society, the scientific societies have not the money to subsidise research. This is done by universities, the Government, industrial firms and endowed bodies. There is no organisation of research on a national scale. Some of the Government and industrial research is secret, and therefore of no value to science. For science means knowledge.

The Royal Society's funds are generally very well spent. So are the Government funds spent through the Medical Research Council, since their use is supervised by a number of committees representing the scientists who will actually carry out the research. Research in universities is uneven. The same professors are supposed to teach and to investigate; and good teachers are often bad investigators, and conversely. They are appointed for life by committees of elderly men, which often choose the wrong man or woman. They are never chosen by the younger workers in the science concerned, who are most competent to discover originality in their colleagues.

The smaller institutions, including hospitals, sometimes do good work, but some are quite inefficient and others definitely corrupt. Even where there is no actual corruption, it is common to find that an old gentleman who attends a committee four times a year is paid twice the salary of a research worker with fifteen years' experience.

Nowhere is research democratically organised. Such a simple device as a wall newspaper would be of great value, provided everyone in a laboratory, from the charwoman to the professor, could use it to state their opinions without fear of the result. Most good scientific work is done by people under forty, and very few people under forty are in charge of laboratories. So the work of a laboratory is often concentrated on problems which interested the professor in his youth, and are now less important than they were.

The British Association is able to spare a few thousand pounds yearly for grants in aid of research. But its main function now is discussion. New results are generally announced at meetings of the smaller societies, and the public hears very little of them. Both in the Soviet Union and in Scandinavia the Press has far better scientific news than in Britain.

If science is to advance in this country as it should, we need more democracy in the laboratories, and also more democratic control of expenditure on research. This will only be possible if the people is educated in science: and they are at present deliberately kept in the dark. For a knowledge of science leads to a realisation of the huge amount of knowledge which could be applied to the public benefit if industry, agriculture and transport were organised for use and not for profit. And knowledge of this kind is dangerous to capitalism.

SYNTHETIC SUPERSTITION

ONE definition of superstition is "Other peoples' religion." I certainly disagree with many points in other people's religious opinions, but I am not going to attack them here.

Religious doctrines, even when they are untrue, are generally part of a system in which some intelligent and thoroughly decent people believe.

I propose to discuss the synthetic superstitions which are being manufactured as "opium for the people," particularly astrology. Large numbers of Sunday newspapers keep a tame astrologer, and I sometimes look at their columns.

Now I don't despise real astrology. It began thousands of years ago in an attempt to link up happenings in the earth with those in the sky. It led to the keeping, in Babylon and other cities of what is now Iraq, of exact records of eclipses and other celestial events which have been of the greatest use to modern astronomers.

And in the late Middle Ages it developed into an art with elaborate rules. In order to cast a horoscope, you had to know the positions of all the planets at the time of a person's birth. Indeed, for accurate prediction the exact hour is needed. Each planet was supposed to have a good, bad or neutral influence.

And the sky was divided up into "houses" concerned with various aspects of human life. Thus if at the time of your birth Saturn and Mars were in conjunction in the House of Death, you were likely to meet with an early and violent death. A given day was or was not lucky for a person whose birth hour was known, according to very complicated rules.

These rules were supposed to embody the wisdom of the ancient Chaldeans. Astrology had a great influence on the thought and language of ordinary people. Such common words as "consider," "disaster," "influence," "influenza," and "conjunction," all derive from astrological theory. Astrology received shattering blows when two large new planets, Uranus and Neptune, were discovered, not to mention one moderate-sized one, Pluto, and about 1,000 dwarf planets. The wise men of the East had never discovered them, and the attempt to fit them into a horoscope is like putting a motor car into a coat of arms.

For astrology, like heraldry, has its rules, and is a quite amusing though rather futile hobby. If I were a genuine astrologer following the great tradition of the "science," I should be even more angry with the Sunday newspaper practitioners than with complete sceptics.

For these ladies and gentlemen predict your lucky days on the basis of the month in which you were born. I was born when the sun was in the constellation called the Scorpion. Now, according to traditional astrology, this alone does not tell me much. If the lucky planets, Venus and Jupiter, were there, too, then I may look forward to certain kinds of good luck.

But, if astrology is true, it is as ridiculous to predict a person's fortune from the position of the sun alone as it would be to diagnose a disease by looking at a patient's tongue without taking his temperature or pulse rate or making any other examination.

Another of these bogus sciences is palmistry, though some palmists certainly manage to size up the character of their clients in a very shrewd manner. But the only palmist whose word I trust as to her methods tells me that she gets most of her information "psychically," i.e. she doesn't quite know how.

There is such a thing as scientific palmistry, founded by a Viennese woman, Hella Pock, who studied the heredity of the folds in the human palm. She found, among other things, that when both parents had a line from the wrist to the middle finger on each hand, 70 per cent. of their children had the same lines, whereas when neither parent had such a line, only 9 per cent. of the children had one on both hands.

If the astrologers and palmists want to convince scientists of the truth of their "sciences," they have an easy task. No doubt (if their claims are right) they must have discovered that millions of young men were going to die between 1914 and 1918. So they ought to be able to predict the dates of future wars. When they get a few such dates right I shall take them seriously. But I am not much impressed by a few lucky shots.

However that may be, astrologers and palmists are very useful to the cause of capitalism. They help to persuade people that their destinies are outside their control. And, of course, this is true as long as enough people believe it. But if enough people learn how the joint fate of us all can be altered, things begin to happen which mean the end of capitalism as well as of astrology and palmistry.

PRACTICE AND THEORY IN SCIENCE

One of the commonest criticisms with which we scientists have to put up is that we change our theories so quickly that they must obviously be worthless. At one time atoms were said to be indivisible; to-day they are split in scores of laboratories. Electricity was first thought to be a fluid, then to consist of particles, and now the particles turn out to behave like groups of waves. Tuberculosis was once said to be hereditary; now it is supposed to be caused by a germ. And so on.

Of course, this is partly due to the fact that science is very badly taught. A scientific theory may be nothing but the truth, but it is never the whole truth. Lenin said that the properties of an electron, the smallest known particle, were inexhaustible—that is to say, there would always be something more to find out about it. So no Leninist should have been surprised when it was found to have previously unexpected properties.

But this is not all. Scientific theories are always guides to practice, or at least to prediction. Chemical theory tells you how to prepare a metal or a drug. Astronomical theory tells you when and where to look for an eclipse. The old theories were certainly wrong. It was supposed that the sun went round the earth, and that when you heated iron ore with charcoal a stuff called "phlogiston" came out of the charcoal and united with the ore to make iron.

Now, we say that the earth goes round the sun, and that oxygen leaves the iron ore to combine with the charcoal (or nowadays coke). But the old theories were partly right. They were right in saying that the sun was further away than the moon, and that the amount of charcoal needed was proportional to the amount of iron to be made.

No doubt our present theories will have to be altered.

But they are truer because they are nearer to practice. One can be sure that one's theory is incomplete and partly wrong, and at the same time be sure that it is near enough to the truth to enable one to do a particular job.

For example, I have eaten about two-thirds of the quantity of ammonium chloride which would kill me. I made some calculations beforehand which were based on the theory that all atoms of chlorine were alike and similarly for nitrogen and hydrogen. This is false. Some hydrogen atoms are twice as heavy as others. But it was true enough, in this connection, for me to stake my life on it.

In the same way, I don't believe in the absolute truth of Marxism in the way that some people believe in religious dogmas. I only believe that it is near enough to the truth to make it worth while betting my life on it as against any rival theories.

Some discarded theories were substantially true when they were first put forward, and ceased to be true later. In the nineteenth century chemists said that atoms could not be split. They tried with all the means at their disposal, such as heat, electric currents, strong acids and alkalies. Those units which resisted their efforts were called "atoms."

Then Rutherford and his pupils developed much more powerful methods, such as protons (nuclei of hydrogen atoms) moving in a million-volt field, and split a number of atoms. The old theory ceased to be true because of these changes in technique.

Just the same happens with political theories. At the

end of the nineteenth century most Marxists thought that socialism could not be achieved in one country alone. They were probably quite right at that time. Then Lenin pointed out that "uneven economic and political development is an absolute law of capitalism. Hence the victory of socialism is possible first in several or even one capitalist country, taken singly." But this only became true when capitalism had developed to the imperialistic stage.

Sometimes two or three apparently contradictory theories are both true. Tuberculosis is due to infection. It is also due to heredity and to bad surroundings. The hereditary element is shown by the fact that, given infection, several members of the same family (and particularly so-called identical twins) will get the disease in the same place, say the base of the left lung or the glands of the neck.

We cannot yet control heredity save in a few cases. But we could see that our children got milk as free from tuberculosis as those of the United States, or that they got the fresh air, sunshine and diet which enable people to fight the infection, even if they have some hereditary tendency to it.

We are quite right to emphasise the environmental factors in tuberculosis just because we can control them. We say that a house caught fire because someone threw down a cigarette end, and not because there is 21 per cent. of oxygen in the air. But it is perfectly true that wood will not burn in air containing only 15 per cent. No event has only one cause. But a refusal to think or act until you know all the causes is not science but pedantry. And a refusal to recognise a new cause when we can control it or even predict its changes is a sign of adherence to obsolete dogma.

So don't worry if we scientists change our theories. It is a healthy sign. "Frankly admitting a mistake," said Lenin, "ascertaining the reasons for it, analysing the conditions which led to it, and thoroughly discussing the means of correcting it—that is the earmark of a serious party." It is also the earmark of a serious scientist.

HISTORY, PHILOSOPHY AND SOCIOLOGY OF SCIENCE

Classics, Staples and Precursors

An Arno Press Collection

Aliotta, [Antonio]. **The Idealistic Reaction Against Science.** 1914

Arago, [Dominique François Jean]. **Historical Eloge of James Watt.** 1839

Bavink, Bernhard. **The Natural Sciences.** 1932

Benjamin, Park. **A History of Electricity.** 1898

Bennett, Jesse Lee. **The Diffusion of Science.** 1942

[Bronfenbrenner], Ornstein, Martha. **The Role of Scientific Societies in the Seventeenth Century.** 1928

Bush, Vannevar. **Endless Horizons.** 1946

Campanella, Thomas. **The Defense of Galileo.** 1937

Carmichael, R. D. **The Logic of Discovery.** 1930

Caullery, Maurice. **French Science and its Principal Discoveries Since the Seventeenth Century.** [1934]

Caullery, Maurice. **Universities and Scientific Life in the United States.** 1922

Debates on the Decline of Science. 1975

de Beer, G. R. **Sir Hans Sloane and the British Museum.** 1953

Dissertations on the Progress of Knowledge. [1824]. 2 vols. in one

Euler, [Leonard]. **Letters of Euler.** 1833. 2 vols. in one

Flint, Robert. **Philosophy as Scientia Scientiarum and a History of Classifications of the Sciences.** 1904

Forke, Alfred. **The World-Conception of the Chinese.** 1925

Frank, Philipp. **Modern Science and its Philosophy.** 1949

The Freedom of Science. 1975

George, William H. **The Scientist in Action.** 1936

Goodfield, G. J. **The Growth of Scientific Physiology.** 1960

Graves, Robert Perceval. **Life of Sir William Rowan Hamilton.** 3 vols. 1882

Haldane, J. B. S. **Science and Everyday Life.** 1940

Hall, Daniel, et al. **The Frustration of Science.** 1935

Halley, Edmond. **Correspondence and Papers of Edmond Halley.** 1932

Jones, Bence. **The Royal Institution.** 1871

Kaplan, Norman. **Science and Society.** 1965

Levy, H. **The Universe of Science.** 1933

Marchant, James. **Alfred Russel Wallace.** 1916

McKie, Douglas and Niels H. de V. Heathcote. **The Discovery of Specific and Latent Heats.** 1935

Montagu, M. F. Ashley. **Studies and Essays in the History of Science and Learning.** [1944]

Morgan, John. **A Discourse Upon the Institution of Medical Schools in America.** 1765

Mottelay, Paul Fleury. **Bibliographical History of Electricity and Magnetism Chronologically Arranged.** 1922

Muir, M. M. Pattison. **A History of Chemical Theories and Laws.** 1907

National Council of American-Soviet Friendship. **Science in Soviet Russia: Papers Presented at Congress of American-Soviet Friendship.** 1944

Needham, Joseph. **A History of Embryology.** 1959

Needham, Joseph and Walter Pagel. **Background to Modern Science.** 1940

Osborn, Henry Fairfield. **From the Greeks to Darwin.** 1929

Partington, J[ames] R[iddick]. **Origins and Development of Applied Chemistry.** 1935

Polanyi, M[ichael]. **The Contempt of Freedom.** 1940

Priestley, Joseph. **Disquisitions Relating to Matter and Spirit.** 1777

Ray, John. **The Correspondence of John Ray.** 1848

Richet, Charles. **The Natural History of a Savant.** 1927

Schuster, Arthur. **The Progress of Physics During 33 Years (1875-1908).** 1911

Science, Internationalism and War. 1975

Selye, Hans. **From Dream to Discovery: On Being a Scientist.** 1964

Singer, Charles. **Studies in the History and Method of Science.** 1917/1921. 2 vols. in one

Smith, Edward. **The Life of Sir Joseph Banks.** 1911

Snow, A. J. **Matter and Gravity in Newton's Physical Philosophy.** 1926

Somerville, Mary. **On the Connexion of the Physical Sciences.** 1846

Thomson, J. J. **Recollections and Reflections.** 1936

Thomson, Thomas. **The History of Chemistry.** 1830/31

Underwood, E. Ashworth. **Science, Medicine and History.** 2 vols. 1953

Visher, Stephen Sargent. **Scientists Starred 1903-1943 in American Men of Science.** 1947

Von Humboldt, Alexander. **Views of Nature: Or Contemplations on the Sublime Phenomena of Creation.** 1850

Von Meyer, Ernst. **A History of Chemistry from Earliest Times to the Present Day.** 1891

Walker, Helen M. **Studies in the History of Statistical Method.** 1929

Watson, David Lindsay. **Scientists Are Human.** 1938

Weld, Charles Richard. **A History of the Royal Society.** 1848. 2 vols. in one

Wilson, George. **The Life of the Honorable Henry Cavendish.** 1851